INTRODUCTION TO GEODESY

Wiley Series in Surveying and Boundary Control

Series Editor: Roy Minnick

INTRODUCTION TO GEODESY

The History and Concepts of Modern Geodesy

JAMES R. SMITH

A Wiley-Interscience Publication

John Wiley & Sons, Inc.

New York / Chichester / Weinheim / Brisbane / Singapore / Toronto

Copyright © 1997 by John Wiley & Sons, Inc.

Library of Congress Cataloging in Publication Data:
Smith, James R. (James Raymond), 1935–
      Introduction to geodesy : the history and concepts of modern
   geodesy / J.R. Smith
         p. cm.—(Wiley series in surveying and boundary control)
      "Wiley-Interscience publication."
      Includes bibliographical references and index.
      ISBN 0-471-16660-X (alk. paper)
      1. Geodesy. I. Title. II. Series.
   QB281.S55   1996
   526'.1—dc20   96-33301

Printed in the United States of America

10 9 8 7 6 5 4 3 2 1

Dedicated to
Eleanor and Peter

Cursed be he who moves his
neighbor's boundary stone.

Contents

Contents

Contents

List of Figures

Preface

To many who study land surveying, whether in its own right or as part of another discipline such as civil engineering, geography, or topographic science, the word *geodesy* can strike an element of fear. This is probably because it is associated with higher mathematics, complicated formulae, and the problems of thinking in three dimensions. To become fully conversant with the subject a high level of mathematical ability is an essential prerequisite.

It should be obvious that to teach such a subject, as with any other branch of surveying, one would have to become involved in all its mathematical intricacies. Here, however, the idea is simply to give the flavor of geodesy and to develop an appreciation of the range of topics encompassed and the many facets of science that it impinges upon. I set out with the idea of trying not to include any formulae or equations and have nearly succeeded. Not until I recently came across an old copy of E. B. Denison's *Astronomy without Mathematics* London 1867 would I have considered it as even a remote possibility.

For those who wish to follow this branch of surveying as a career the daunting elements have to be firmly grasped, but on the other hand it is quite feasible to get an overall feel for the discipline without becoming involved in the intricacies. It is the aim here, as it was in *Geodesy for the Layman* and *Basic Geodesy*, which this replaces, to put across the elements of geodesy in an elementary way, using simple language and with the sid of numerous diagrams.

The pure geodesist may well have criticisms, since in order to achieve simplicity I have made generalizations that in the strict sense may not be correct but which are sufficiently near the truth for the purposes of this volume. Certainly any reader with more than a passing interest must refer to specialized up-to-date texts. Good examples are *Geodesy* by Guy Bomford and *GPS Satellite Surveying* by Alfred Leick.

Geodesy is a very complicated, highly mathematical science, and in no way should readers assume that this text is any more than an outline introduction. It covers the scope of geodesy and its capabilities but does not attempt to touch many of the intricate areas that have to be accounted for in rigorous applications. For example, some explanation is given of ocean tides but not of earth tides, and of equipotential but not of the spherical harmonics that can be used to express it. On the other hand, this new edition does include brief mention of some of the more abstruse quantities, such as nutation and aberration.

Many instruments are mentioned, but without details as to the necessary adjustment and observational procedures and subsequent data analysis through statistical techniques. Variations in atmospheric conditions play a considerable role and are still far from adequately understood.

Particularly as we enter the satellite-reliant age, a wider range of professionals will come into contact with geodesy and require an outline knowledge of its capabilities and fields of application.

The area of satellite research and the general application of electronics to surveying is changing so rapidly that it is difficult to keep pace with developments. Here no attempt has been made to be up to date immediately with technological developments. Rather, an outline is given of basic techniques and principles. In many instances these survive equipment changes with only slight modification.

Much of the material in this book is based on the two very popular predecessor volumes mentioned above, with updating especially in areas of rapid change, such as those involving satellites. However, in the short 8 years since the first edition of *Basic Geodesy*, the use of satellites for positioning has progressed so rapidly on a global scale that it now warrants a much enlarged, separate treatment. Thus, the present volume should not be taken as a definitive textbook on GPS; it is an introduction intended no more than to lay the foundations. It must be said that many readers of the first edition have suggested that more space should be given to GPS, but this volume is really meant to focus on the basics of geodesy. Hence a fine line is being trod between maintaining the theme and trying to satisfy potential readers. On the other hand, it can truthfully be said that with the introduction of satellites new factors influence measurements for geodetic and these are mentioned in the appropriate sections.

An additional chapter has been added in this new edition to illustrate the wide range of modern projects where geodesy in one form or another is very pertinent. The examples are by no means exhaustive but represent a cross section of topics that are fully written up elsewhere. The interested reader can delve deeper as he or she so wishes.

Previous to the 1950s no one in geodesy was interested in more than the first few hundred meters of the atmosphere or in the characteristics of electromagnetic wave propagation. If stations were not intervisible, they were not readily of use, and thus hilltops were a priority. Now that is no longer the case.

Above all, it is my wish for this volume to be a painless initiation into geodesy and a precursor to immersion in textbooks for the serious student or, alternatively, to provide a general understanding for those in cognate professions. A selected, but by no means comprehensive, bibliography for further reading is included at the end of the book.

Particular thanks must be extended to the original authors for the clarity of both their text and diagrams. They have served the profession well for forty years, and it is to be hoped that this book, in its small way, will be able to match their success. Among colleagues to whom thanks must be extended

I mention in particular Alan Wright of Global Surveys Ltd., who has had considerable input on both the first and second editions, and Carl Calvert, Adam Chrzanowski, Kevin Dixon, Dave Doyle, Kurt Egger, Keith Greggor, Alan Haugh, and Walt Robillard. Thanks must also be extended to Roy Minnick for setting the challenge of updating such a popular works.

J. R. SMITH
Petersfield, U.K. 1997

Second edition

INTRODUCTION TO GEODESY

Introduction

What is this subject of geodesy? Is it just an esoteric area of science with little modern practical application, or does it function quietly to the benefit of all but receive little public attention? Is it a product of the age of electronics, or does it have a much longer history? In brief, what is it, how has it developed, what areas of life does it impinge upon, and how is it put to practical use?

Definitions vary from the inevitable one line "science of measuring the earth, or surveying any large part of it" to detailed paragraphs setting out all the implications of that single sentence. It would serve no useful purpose to offer yet another definition solely for this volume. Instead, the following is an amalgam of several current definitions so you can appreciate from the start the range of topics involved.

Geodesy, from the Greek, literally means dividing the earth, and as a first objective the practice of geodesy should provide an accurate framework for the control of *national* topographical surveys. Thus geodesy is the science that determines the figure of the earth and the interrelation of selected points on its surface by either direct or indirect techniques. These characteristics further makes it a branch of applied mathematics, one that must include observations that can be used to determine the size and shape of the earth and the definition of coordinate systems for three-dimensional positioning; the variation of phenomena near to or on the surface, such as gravity, tides, earth rotation, crustal movement, and deflection of the plumb line; together with units of measurement and methods of representing the curved earth surface on a flat sheet of paper.

Unfortunately there is a trend today for the term *geodesy* to be applied in an umbrella manner, particularly in the European Community, to describe all activities from valuation, land management, soil testing, cartography, setting-out, underground surveys, national mapping, boundary surveys, Land Information Systems, and in fact every activity except geodesy in its traditional definition! In addition, we are now being pressed to use the term *geomatics* to cover almost as wide a selection as the above list. Maybe it will soon be possible to refer to geomatics and geodesy as covering everything that is understood to be under the authority of a surveyor.

Some 65 000 geodetic surveyors are said to work in the European Community (EC) alone! At least two noughts should be deleted from this figure to get to the traditional geodesists and those who fit the sphere of activity of the International Association of Geodesy (IAG)—an organization that would recognize few, if any, of the topics encompassed by the EC statement.

The problem, unfortunately is further complicated by the wide range of uses now being found for the global positioning system (GPS). It is an earth-centered system, relying on earth-orbiting satellites, but it has recreational uses in addition to survey applications and acceptable requirements from a few centimeters to many meters depending on the use.

To quote from the *ACSM Bulletin* of March/April 1992:

> Is the ability to routinely determine boundary lines accurate to $1:100\,000$ by GPS within the province of the land surveyor? I think not unless that person is also trained in the geodetic sciences.
>
> GPS is a tool of geodesy—land surveyors using GPS are practicing geodesy. Therefore they should be thoroughly trained in the geodetic science. This includes higher mathematics, physics, measurement theory, analysis, adjustment computations, geodetic concepts and field experience. (Bloodgood, 1992)

Here it is the intention to follow more the view of the IAG than that of the EC, but with some editorial license. To be of wide topical use, this second edition has a section on GPS that is perhaps disproportionately long. No apology is given for this, as it is a current tool of geodesy and as such warrants fuller treatment than the more historic topics.

In this last quarter of the 20th century so much now revolves around very accurate positioning—whether for oil rigs many kilometers offshore, intercontinental missile targets, shuttle missions into space, monitoring crustal movements, laser bathymetry, navigation, seismic surveys, gravity observations, deformation measurements, all forms of military positioning, or small-scale mapping from space—that the fruits of geodetic labors have an almost limitless range of applications.

However, before venturing into the realms of artificial satellites it will be instructive to trace the origins of geodesy from the first few centuries B.C. up to the present day. From the times of a flat earth concept, through the sphere and spheroid to the geoid; from the knotted rope for measurement to suspended wires, electromagnetic systems, laser ranging to the moon, and the use of orbiting satellites.

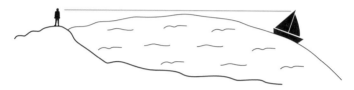

(a) Ship Disappearing over Horizon

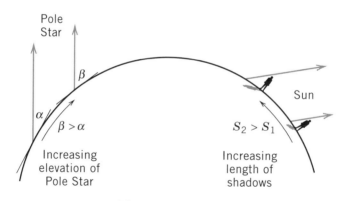

Pole
Star

β

α

$\beta > \alpha$

Increasing
elevation of
Pole Star

Sun

$S_2 > S_1$

Increasing
length of
shadows

(b) Travelling North

1. Realizing the Earth Is Not Flat

CHAPTER
1

History of Geodesy

From earliest times man was increasing his knowledge of the planet on which he lived, although much of its function would have been a mystery. As he tilled the fields, he noticed the changing lengths of daylight and the changes in the maximum height of the sun as the year progressed. If he happened to be a fisherman or simply a watcher of seaborne activity, he could well have realized that when a distant vessel disappears from view the lower part goes first and the top of the mast is the last to vanish. If a person stands on the beach (figure 1a) with his eyes 2 m (meters) above the water level, and the top of the mast of a retreating vessel is 30 m above water level, the vessel would first begin to disappear when 5½ km (kilometer) from shore and finally vanish when 22 km out—not that humans had the means at that time either to measure the distance or to use kilometers!

As man became more adventurous and traveled farther from his home territory, he would have appreciated that if he went in an easterly or westerly direction, the star now recognized as the Pole Star, and around which the signs of the zodiac revolved, would stay more or less at the same altitude, whereas if he went in a northerly or southerly direction, the altitude would change (figure 1b).

In addition, he may well have noticed that under the heat of the midday sun the length of his shadow changed as he traveled north or south, whereas, for east or west travel over periods of a few days, the shadows remained sensibly constant in length.

Whether or not these phenomena gave him any early ideas as to the shape of the earth he was standing on is unknown. One might even question whether the shape of the earth was of any interest at all since travel was very limited and adventurers few and far between. However, thoughts on its shape gradually progressed from flat, through a disc or short cylinder, to a number of variations on these.

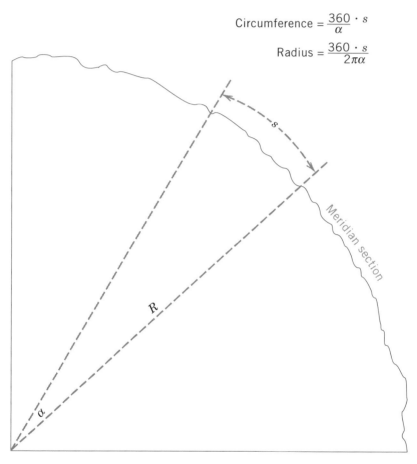

$$\text{Circumference} = \frac{360 \cdot s}{\alpha}$$

$$\text{Radius} = \frac{360 \cdot s}{2\pi\alpha}$$

2. Principle of Determining Size of Earth as a Sphere

PYTHAGORAS

Certainly by the time of Pythagoras (c. 580–500 B.C.) the earth was considered to be spherical—if for no other reason than that the sphere was thought by the philosophers of the time to be the perfect regular solid. As might be expected, not all authorities agreed, and even as late as the sixth century A.D., scorn was still being poured on the idea of a round earth and queries raised as to how such a shape could retain the waters of lakes and oceans.

If we had a table tennis ball, there would be no difficulty measuring its diameter with calipers. Enlarge this to a dome 1 km across, and, while the problem would be a little more difficult, there would still be several ways of measuring it. We might use techniques that were the reverse of setting out curves—for example, we could directly measure the tangent lengths or calculate offsets from a tangent and then do the necessary calculation.

But what happens when the dome has a radius of some 7000 km? Some totally different approach is required. Granted, in the light of methods available today, there would be no difficulty in using satellites, but what about 2000+ years ago, before the developments of the electronic age?

ARISTOTLE

First attempts at putting a dimension on the sphere are credited to Aristotle (c. 384–322 B.C.) who recorded a diameter of 400 000 *stades*. This figure could vary from 84 000 to 63 000 km, depending on our choice of conversion factor. But how he arrived at his value is unknown.

ARCHIMEDES

A century later, Archimedes quoted 300 000 stades (63 000 to 47 000 km). It might even be that each used a different length of stade and that both used the figures equivalent to 63 000 km—possibly from the same initial guess.

ERATOSTHENES

For an approach of any scientific significance we turn to Egypt, where the Greek philosopher Eratosthenes (276–195 B.C.) was librarian at the famous library in Alexandria. He adopted a principle, shown in figure 2, that is still acceptable today. If for the moment we assume the earth is a sphere, then its

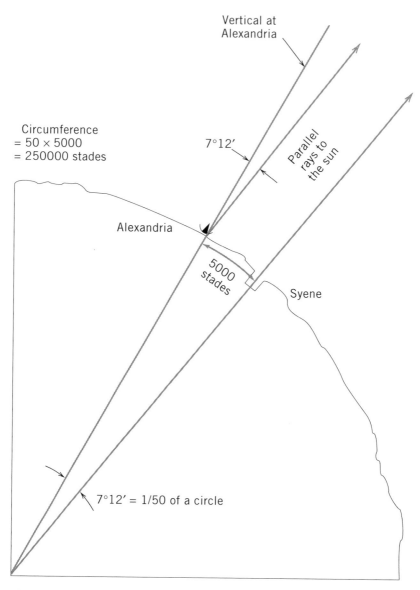

Vertical at
Alexandria

Circumference
= 50 × 5000
= 250000 stades

7°12′

Parallel
rays to
the sun

Alexandria

5000
stades

Syene

7°12′ = 1/50 of a circle

3. Method of Eratosthenes for Determining Size of the Earth, 230 B.C.

size can be found if two quantities are known: the distance s between the two points that lie on the same *meridian* (a line joining the two poles), and the angle α subtended by those two points at the center of the earth. Thus

$$\text{circumference} = 360° \cdot \frac{s}{\alpha}.$$

On the face of it the determination of α might appear particularly difficult, but luckily there is no need to get to the center of the earth to observe its value. This can be achieved on the surface in various ways, as illustrated in the following examples.

In his position as librarian, Eratosthenes had access to a vast range of material and ideas. Thus, whether he actually made the measure with which he is credited or it was a combination from other reports is open to question.

Nevertheless, as the story goes (figure 3), Eratosthenes noticed that at the summer solstice the sun shone directly into a deep well at Syene (Aswan). This spot would have to be on the tropic of Cancer for this observation to apply. At the same time at Alexandria he found (probably by using a scaphe—a hemispherical bowl with a central gnomon) that the sun cast a shadow equivalent to ⅟₅₀ of a circle—about ¼ of a sign of the zodiac—or $7°12'$ in present terminology. This angle is the same as the one subtended at the center of the earth by the two terminal points on the earth's surface, Syene and Alexandria. He combined this measure with a value of 5000 stades for the distance between the two cities. Whether this was measured by the royal road surveyors, estimated in terms of camel-days journey, or an educated guess is unknown.

The observations and assumptions are full of possible errors. The modern equivalent of $5000 \times 50 = 250\,000$ stades (52 500 to 39 400 km) is close to the present-day accepted distance of around 40 000 km, but this can only be considered fortuitous, for several reasons. Syene is not on the tropic of Cancer but some 60 km north of it; Syene and Alexandria are several degrees away from being on the same meridian, and the measure of their distance apart was some 10% in error. Then when we realize that Eratosthenes apparently added 2000 stades simply to make the result divisible by 60, the uncertainty of the whole calculation becomes apparent.

Aside from errors of position, varying interpretations of the stade, and other such problems, what about basic inaccuracies in measuring the distance and the angle? Let us assume for the moment that the earth is a perfect sphere (it makes the arithmetic far simpler), an arc of $1°$ is measured, and the distance between the ends is found to be X meters (about 110 000 m). Then the circumference is $360X$ meters. If the angle was incorrectly measured by $1'$ of arc, the resulting circumference would be incorrect by $6X$ meters (approximately 660 km). Conversely, if the distance were incorrect by 1000 m, then the circumference would be incorrect by 360 000 m (360 km). Neither of these figures is insignificant, and yet both would have been possible with the equipment available at the time.

Circumference
= 48 × 5000
= 240000 stades

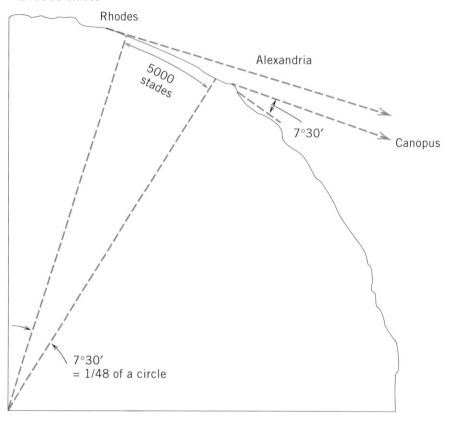

4. Method of Poseidonius for Determining Size of Earth, 100 B.C.

POSEIDONIUS

A century later, Poseidonius used a different technique (figure 4) to obtain a comparable result of 240 000 stades. He noticed that the star Canopus was on the horizon when viewed from Rhodes while at Alexandria its elevation was ⅛₈ of a circle (7°30′). The distance between the two cities was estimated as 5000 stades, but since they are separated by water, this could only have resulted from mariners' estimates.

As with Eratosthenes, the angle used by Poseidonius was considerably in error, in fact by 2¼°, and the distance was off by some 30% to 40%. The errors were compensatory, however: the too-small angle and too-long distance combined to give a fortuitously acceptable result.

Difference in shadow lengths is
a function of angle α

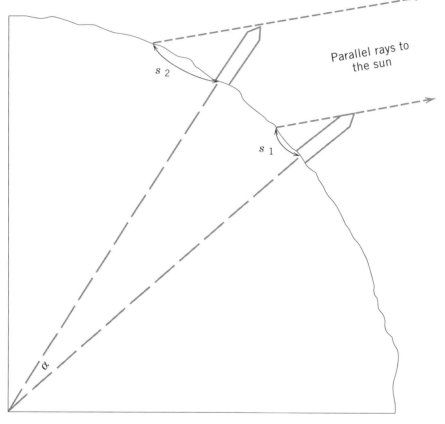

5. Method of I-Hsing for Determining Size of Earth, A.D. 724

I-HSING

Another variation on the same theme was used in China in the eighth century A.D. (figure 5) when I-Hsing, a Tantric Buddhist monk who was also a famous mathematician and astronomer, was charged with organizing observations over an arc of some 11 400 li (5000 km) on the 114° east meridian. This requirement arose from the long-held belief that the shadow length of an 8-foot (1.96-m) gnomon changed by 1 Chinese inch (0.02 to 0.03 m) for every 1000 li traveled along a meridian. In the event, the rate of change was more nearly 4 inches (0.1 m). The same data, however, gave a value for the earth's circumference of 128 300 li (56 700 km).

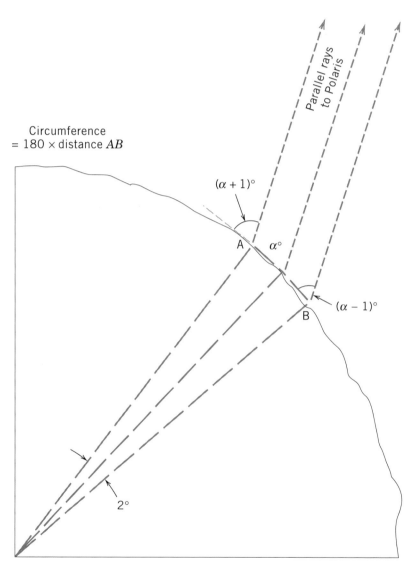

Circumference
= 180 × distance AB

$(\alpha + 1)°$

A

$\alpha°$

$(\alpha - 1)°$

B

Parallel rays
to Polaris

2°

6. Method of Al -Mamun for Determining Size of Earth, A.D. 820

AL-MAMUN

Yet another alternative was given in A.D. 820 by the Caliph Al-Mamun in Arabia (figure 6). He directed his astronomers to measure several (probably 4) lines around Baghdad and Al Raqqah. From a central starting point, parties went both north and south until the vertical angles to Polaris (the Pole Star) had changed by 1°. How the distances were measured is again open to doubt. Some records suggest long knotted ropes, others the distance traveled by horsemen in a given time. However measured, the distances were recorded in Arab miles with an accepted result of 56⅔ Arab miles for 1°. Conversion is again uncertain, but an accepted equivalent of 111.073 km gives the circumference as 39 986 km.

COLUMBUS

It was probably such uncertainties as these for the size of the earth that led Christopher Columbus to think that Asia was only some 4000 miles (6400 km) west of Europe. For example, the geographer to whom all seafarers of the Middle Ages referred was Ptolemy (A.D. 100–178), who accepted a value for the earth's circumference of 180 000 stades. This could be converted by the wrong factor to 28 350 km. If this were then used to determine the circumference at, say, 36°N latitude, the result would be 23 200 km, as compared to the more realistic value of about 32 800 km. If the earth were thought to be only some 70% of its actual size, then the navigator/explorer had little chance of success. Some historians give an alternative explanation that Columbus used the result of Al-Mamun but assumed it was given in Italian (Roman) rather than Arab miles—a difference of some 25% too small.

Yet another way of accounting for the shortfall in Columbus's estimate is to consider that a degree of latitude was taken to be 56⅔ Roman miles instead of the more nearly correct 75 Roman miles. This results in a ratio of 0.76, again producing a figure far too small.

It is not my intention here to examine the Columbus voyages in detail. In summary, however, one scenario might have been that Columbus took the inhabitable world to be 13 750 Roman miles and 300° extent at 36° N, whereas it should have been more like 9600 miles and 160°. For the earth's circumference he took 16 500 instead of 21 580 miles. Thus sailing west from Europe using his figures would require traveling only 60° longitude and 2750 miles, whereas it should have been 220° and 12 000 miles. Many other scenarios could be set up, but they would all imply that the distance going west was far shorter when in fact it was far longer. It is likely that Columbus not only underestimated the size of the earth but also overestimated the size of the area known as the inhabitable world.

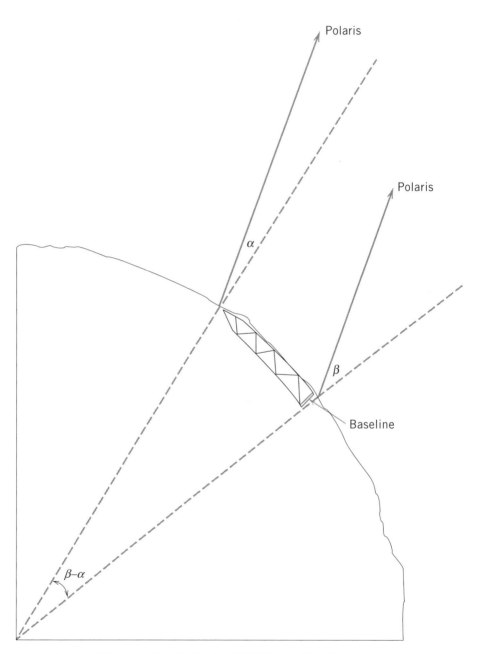

7. Use of Triangulation by Frisius (1533), Snellius (1620s)

The 16th and 17th centuries were the turning point in most aspects of science in the Western world, and this applies also in no small measure to surveying. Developments in instrumentation—including the telescope, vernier, thermometer, and barometer—and in computing techniques such as logarithmic and trigonometric tables all opened the way for the introduction and development of triangulation. (See Chapter 4.)

FRISIUS AND SNELLIUS

In a publication of 1533, Gemma Frisius put forward the principles of triangulation, but it is uncertain whether he actually put them into practice. The scheme he illustrated was certainly not observable. The technique was further developed by Tycho Brahe in the late 16th century, but the real landmark was the technique used by Willibrord Snellius in the 1620s.

Between Bergen-op-Zoom and Alkmaar in the Netherlands, Snellius observed a triangulation scheme with five baselines in the vicinity of Leiden. This technique deviated from all previous arc measures in that the distance between the terminal points could now be determined indirectly rather than directly. That is, instead of measuring 100+ km by tape or similar device, all that was now required was to very accurately measure at least one line (in the case of Snellius the lines averaged 1300 m) for the calculation of an arc of around 130 km. (See figure 7; triangulation is explained in more detail in Chapter 4.)

FERNEL

At about 1525, Jean Fernel measured an arc of 1° between Paris and Amiens using the revolutions of a carriage wheel to determine the distance. For his angle he first measured the height of the sun at midday. He then calculated what the height would be at a point 1° farther north and traveled in that direction until he found a position where he was able to observe that value. (There was a small error in his calculation, but that is only to be expected.)

He found the distance between these two points to be 68.096 Italian miles. There is some doubt about the exact equivalent of the Italian mile in other units such as the *toise*, although it is generally taken as 56 746 toise.

PICARD

In 1669, l'Abbé Jean Picard introduced the telescope for observing star altitudes and angles of triangulation. At that point the observation work became similar in principle to that subsequently used until the last quarter of the 20th century, with the advent of the artificial satellite and the global positioning system (GPS). With wooden rods, Picard measured two baselines of 5663 and 3902 toises (11 038 and 7605 m) on his triangulation from Malvoisin near Paris to Sourdon near Amiens. He calculated that 1° equaled 57060 toises (111 210 m).

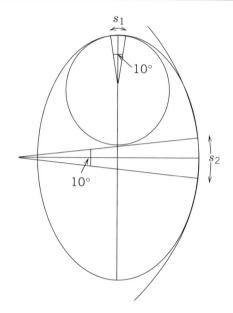

(a) Prolate $s_2 > s_1$

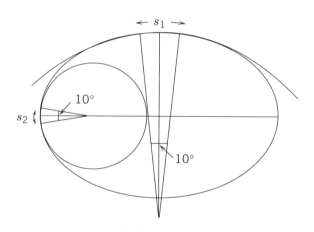

(b) Oblate $s_1 > s_2$

8. Oblate and Prolate Spheroids

THE CASSINIS

By the end of the 16th and early 17th centuries, several conflicting measurements and calculations as to the shape of the earth had been made in Europe. These in turn affected estimates of the earth's size.

Contemporaneous with the founding of the Royal Society in London and l'Académie Royale des Sciences in Paris came European expeditions abroad. These had wide terms of reference, such as observations of the oscillations of a pendulum at different elevations and in different latitudes, and of changes in the velocity of sound and of atmospheric pressure with elevation. These observations, together with the theories of Isaac Newton, suggested the earth must be flattened at the poles.

Almost in parallel with these activities, several long arcs were being measured in France under the guidance of the Cassini family, who occupied posts in the Paris Observatory over four generations. All the French arc measurements suggested equatorial flattening rather than flattening at the poles. How could such overwhelming evidence be disputed by theory? As we shall see, the discrepancies resulted from the inadequacy of the available instrumentation, which was unable to detect the small changes involved.

This was a scientific controversy par excellence—as it has been described by some, the battle between the pumpkin and the egg shape. The technical names given to the rival shapes are *oblate spheroid* for flattening at the poles and *prolate spheroid* for flattening at the equator (figure 8). For any given angular value, the equivalent arc length will increase toward the equator for a prolate spheroid and increase toward the poles for an oblate one.

It will be noticed from figure 8 that when the earth is no longer considered to be a sphere then the varying radii of curvature of the surface do not meet at a single central point.

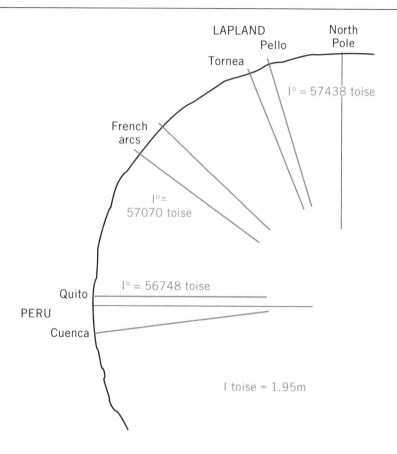

9. Spread of 18th-Century Arcs

The controversy over shape needed urgent resolution, since it had critical implications for the growing number of intrepid explorers and navigators, as witnessed earlier in the case of Columbus.

To resolve the debate, in the early 1730s it was suggested to the Academy in Paris that an expedition be sent to as near the equator as possible to measure a long arc and a second expedition should go as near to the North Pole as possible to do the same. Two such widely separated arcs (figure 9) should prove conclusive: if the length of 1° of arc was greater near the North Pole than it was near the equator, then Newton would be proved correct; if the opposite, then the Cassinis would be right.

So it was that groups of Academicians went to Peru (now Ecuador) and to the borders of Sweden and Finland in southern Lapland. Several famous people were among these parties: Pierre Bouguer and Charles-Marie de La Condamine went to Peru in a party that included two Spanish naval officers; Pierre-Louis de Maupertuis, Alexis-Claude Clairaut, and Anders Celsius went to Lapland.

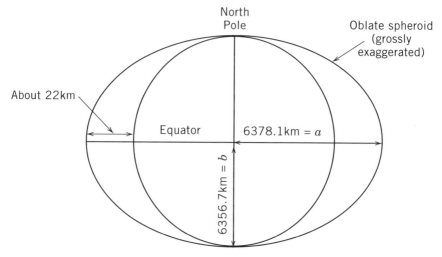

North
Pole

Oblate spheroid
(grossly
exaggerated)

About 22km

Equator

$6378.1\text{km} = a$

$6356.7\text{km} = b$

10. Relation of Oblate Spheroid to Fitting Sphere

Their results convincingly proved Sir Isaac Newton to be correct and the Cassinis wrong. As Voltaire said at the time about Maupertuis, "Il avait aplati la Terre et les Cassinis." Maupertuis in a letter to the British astronomer James Bradley wrote, "Thus Sir, You See the earth is Oblate, according to the Actual Measurements, as it has been already found by the Law of Staticks: and this flatness appears even more considerable than Sir Isaac Newton thought it." Later Voltaire commented to returning members of the Peruvian expedition, "You have found by prolonged toil, what Newton had found without even leaving his home."

The two main radii of the earth calculated from the results of these expeditions were 6376.45 and 6355.88 km. In terms of the length of 1° of arc, this gives a difference of only 1350 m in 111 km. It is not surprising that the errors inherent in the equipment and techniques of the Cassinis swamped the small amount they were trying to resolve. Figure 10 shows more recent, but only slightly different, values and indicates the relationship to a fitting sphere.

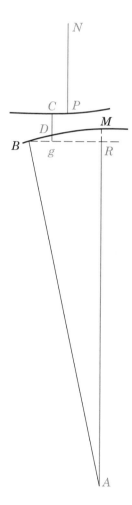

11. Mason and Dixon's (1766) Method
of Arc Measure

MASON AND DIXON

To illustrate the complicated nature of comparisons between arcs because of different units of measure, let us consider the arc in the United States measured in 1768 by Charles Mason and Jeremiah Dixon. (This is *not* the famous boundary line between Pennsylvania and Maryland, but the arc of the meridian, which they measured at the same time.)

They found 1° to be 363 763 English feet at a mean latitude of 39°12′. Now all other arcs for comparison were in French toises. Thus, to quote from Charles Huttons' Abridged Philosophical Transactions,

> To reduce the measure of a degree to the measure of the French toise, it must be premised, that the measure of the French foot was found on a very accurate comparison made by Mr Graham, of the toise of the Royal Academy of Sciences at Paris, with the Royal Society's brass standard, to be to the English foot as 114 to 107. . . . , which, divided by six, the number of feet in a toise, gives the length of the degree as = 56 904½ Paris toises. . . . Such is the length of a degree in this latitude supposing the 5 ft brass standard made use of in the measure to have been exactly adjusted to the length of the Royal Society brass standard. It was really adjusted by Mr Bird by his accurate brass scale of equal parts . . . and which is just 1/1000 th part of an inch shorter than the Royal Society's standard. . . . He then finds other reasons to amend the result by a further 3 feet or a total of 13 feet to be subtracted. But because of other considerations that would negate this correction he concludes that the best value should be 56 904½ toises!!

What all this illustrates is that the smallest uncertainty in a standard of measure, when multiplied up to the length of a degree, can be very significant. The figure of 56 904½ comes from 363 763 × (107/114) × ⅙. But if Bird's scale was shorter by 1/1000 inch over a yard, or 1/36 000, then the whole arc would be affected by about 10 ft (1½ toises).

The arc Mason and Dixon used was not of a design comparable to other arc measures, either of the same vintage or any other time, so it is worth a brief description here.

In terms of figure 11 it was required to determine the length and amplitude of line *NA*. Mason and Dixon first traced the meridian from *N* to *P*, a distance of 14.8 miles (23.8 km). The point *C* was then placed in the same latitude as *P* at a distance of 2.99 miles (4.8 km). The meridian was then traced from *C* to *D* to get 5.0 miles (8.0 km). Point *B* was then found such that *BDM* was a straight line with *BD* = 0.28 mile (0.45 km) and *B* was south of *D* by 90 ft (27.4 m). *BA* was then traced as 81.98 miles (131.9 km). This last length was then moved proportionately to arrive at *AR*. *Thus*

$$NA = AR + gD + DC + PN$$
$$= 538\ 067 \text{ ft}$$
$$= 1°28′44.9″$$

and

$$1° = 363\ 763 \text{ English feet.}$$

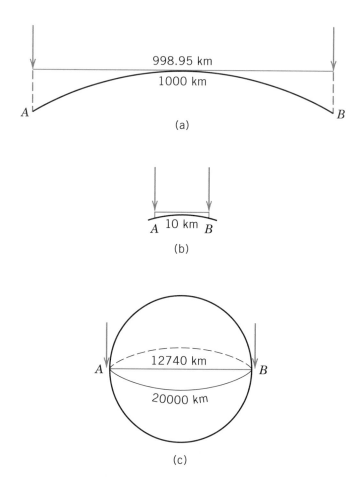

998.95 km

1000 km

A

B

(a)

A 10 km B

(b)

A 12740 km B

20000 km

(c)

12. Curved Surface and Plane Equivalent

CHAPTER
2

Earth's Shape

The previous chapter has recounted how geodesy was born, but what is all this talk about spheroids—surely all can see that the surface of the earth is highly irregular in shape?

While the irregularity is an indisputable fact, there are severe difficulties in even attempting any mathematical calculations on such a surface. For practical purposes it is necessary to find a regular solid figure that most nearly matches the topographic surface of the earth. Remember that even the most rugged topography of the Himalayas, at some 8 km high, is only a thin veneer on an earth of nearly 6400 km radius. If the topography of a globe of 2-m radius were shown to scale, it would be no more than 2½ mm thick.

The oblate spheroid mentioned earlier is a refinement of the true sphere such that its N-S radii are slightly less than those in the E-W direction at the equator. The difference between the two axes is only some 22 km, so again it is difficult to show even this correct to scale. On the 2-m-radius globe the distortion would only amount to 7 mm, which would not be discernible to the naked eye. Thus it is that all diagrams (such as figure 10) have to be grossly distorted to illustrate the situation.

Even so, the topography does complicate the issue because of its irregularities. Those parts of the earth that have the most regular surfaces are the oceans. If we imagine them extending under the continents in small frictionless channels, then the surface represented might be useful. In fact such a surface would be close to the accepted oblate spheroid—and will be met again.

Following this idea, the mean sea level is taken as a reference surface, and any measurements on the irregular topography are reduced to their equivalents on this sea-level surface (or one near it).

But this leads to the further question of what is meant by "sea level," as we are all aware of the rise and fall of the sea as we sit on the beach in the summer sun (Chapter 5 will consider the sea-level problem further.)

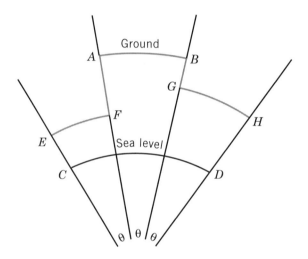

13. Effect of Elevation on Arc Length

Now with an earth of such shape and dimensions, if we are interested in the relation of widely separated points, we need to take the shape into consideration. On the other hand if only small areas are of interest, it may well be feasible to revert to considering the earth flat. But what is meant by "small area"? This is not readily defined; it will depend upon such factors as what we wish to depict and the degree of accuracy we seek in the answer.

What might be the criteria, and can they be varied at all? These questions lead us into the field of map projections (Chapter 15). However, for the moment, to get a feel for the situation, consider two points A and B at the extremes of an area of interest and 1000 km apart (figure 12a). To consider part of the earth flat we could in effect imagine a sheet of paper touching the globe in the vicinity of the area of interest. If it were possible to look at this situation from outer space (that is, from near infinity), the two points would appear to be 998.95 km apart, whereas their true separation would be 1000 km, a distortion of more than 1 in 1000. If this were taken to the extremes of

1. A and B only 10 km apart (figure 12b), then the difference is a few millimeters.

2. A and B as points at opposite ends of a diameter at the equator (figure 12c), then the apparent distance AB would be only ⅔ of the true distance.

But what of the effect of elevation above mean sea level? In figure 13 imagine three plateau areas each of which subtend the same angle θ at the center of the earth but which are at different elevations. One of the many positions where the three distances are represented as being the same length (since the angles θ are the same) is their equivalent representation at sea level. For all mapping purposes this type of reduction is used.

If the line AB is 2000 m long and 1000 m above sea level, the equivalent distance at sea level would be 1999.68 m. While the difference in this example is relatively small, but not insignificant, it can obviously become large over long distances.

Although by the 1730s the earth was recognized to have a flattened form, refinements have continued to be made to get the best values for the parameters of the spheroid. Since the Peru and Lapland arcs there has been a steady stream of scientists, astronomers, and surveyors who have analyzed various observations to produce new values for the parameters and so get their names into the annals of survey history. Each successive reevaluation has been named after the relevant personage, and some, like Everest and Clarke, have several claims.

Now, whenever a country has decided to carry out national survey work, it has selected a particular set of earth parameters for the computation. While usually it has chosen the best value known at the time, or maybe one dictated by a national determination, subsequent parameters are probably much better.

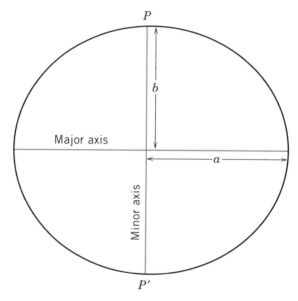

a = semi major axis
b = semi minor axis
F = flattening = $\dfrac{a-b}{a}$
PP' = axis of revolution of the earth ellipsoid

14. Elements of an Ellipse

However, once a choice has been made, so much is then dependent on it that later changes to better parameters are unlikely.

For example, Great Britain was active in national survey work in the first half of the 19th century and so selected the values of their astronomer royal, Sir George Airy. Subsequent work has indicated that the choice was not a good one; nevertheless it continues to be used since it is now the basis of all of national mapping.

EARTH PARAMETERS

The parameters by which a spheroid (ellipsoid) is defined (figure 14) are the values of the two main axes, where

a = the semimajor axis or equatorial radius,

b = the semiminor axis or polar radius.

Alternatively the figure is fully defined by a and the relationship $(a - b)/a$, which is called the *flattening* and denoted by f. All modern calculations place the value of f close to $1/300$.

The following list gives a selection from the numerous values available and indicates some of the areas in which they have been adopted.

Year	Name	a(m)	b(m)	$1/f$	Where used
1830	Airy	6 377 563	6 356 257	299.325	Great Britain
1830	Everest	6 377 276	6 356 075	300.802	India, Pakistan, Burma
1841	Bessel	6 377 397	6 356 079	299.153	Germany, Indonesia, Netherlands, Japan, N.E. China
1858	Clarke	6 378 294	6 356 618	294.261	Australia
1866	Clarke	6 378 206	6 356 584	294.978	USA
1880	Clarke	6 378 249	6 356 515	293.466	Cen., S., and W. Africa; France
1907	Helmert	6 378 200	6 356 818	298.300	
1909	Hayford	6 378 388	6 356 912	297.000	USA; from 1924 used internationally; Europe, N. Africa
1927	NAD 27	6 378 206.4		294.978 698 2	
1948	Krassovsky	6 378 245	6 356 863	298.300	Russia, Eastern countries
1960	Fischer	6 378 155	6 356 773	298.3	S. Asia
1966	WGS 66	6 378 145	6 356 760	298.25	
1967	IUGG	6 378 160	6 356 775	298.247	Internationally; W. Europe, S. America, Australia
1972	WGS 72	6 378 135	6 356 751	298.26	From satellite geodesy
1980	International	6 378 137	6 356 752	298.257	Recent best fit to geoid
1983	NAD 83	6 378 137.0		298.257 222 101	
1984	WGS 84	6 378 137		298.257 223 563	

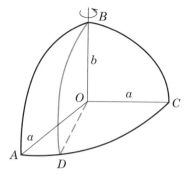

(a) Oblate Spheroid

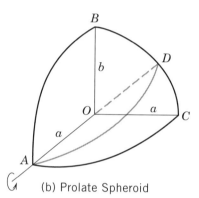

(b) Prolate Spheroid

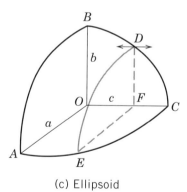

(c) Ellipsoid

15. Spheroids and Ellipsoid Defined

It is noticeable that, whereas only 50 years ago new determinations could differ from previous ones by hundreds of meters, today the modifications are of the order of 1 to 2 m only.

Before accepting any of the above values for practical application, the user should contact the appropriate documentation office of a country so as to ascertain any local modifications to the numbers.

SPHEROID AND ELLIPSOID

Two confusing terms will occur in the context of the earth figure—spheroid and ellipsoid. Is there any difference? Certainly they tend to be used interchangeably. Mathematical reference books define them separately: a *spheroid* is a solid generated by rotating an ellipse about either axis, while an *ellipsoid* is a solid for which all plane sections through one axis are ellipses and through the other are ellipses or circles. However, if any two of the three axes of the ellipsoid are equal, the figure becomes a *spheroid,* and if all three are equal, it becomes a *sphere.* In the case where two of the axes are equal, the strict definition is then of an *ellipsoid of revolution.*

The dictionary suggests, however, that the term *spheroid* can also refer to any slightly nonspherical shape but not necessarily a mathematically definable one.

If an ellipse is rotated about its minor axis, it produces an oblate shape, and if rotated about its major axis, it produces a prolate shape (figure 15).

Thus in figure 15, for oblate there is rotation of *BDO* about *BO* and for prolate rotation of *ADO* about *AO*. In the ellipsoid, when *EDF* moves parallel to *AOB*, it produces ellipses of varying size.

Though modern texts use the terms in an arbitrary manner, in geodetic circles there is a tendency toward accepting the terms *ellipsoid* and *ellipsoid of revolution.*

As you can see from the table on page 31, there are many ellipsoids. The first international figure was developed in 1924, but today the advanced technology of GPS and related systems allows the use of various world-based figures (see Chapter 10). Even so, the most acceptable figure, DOD WGS 84, differs from the geoid (see next section) in the range of +60 to −100 m.

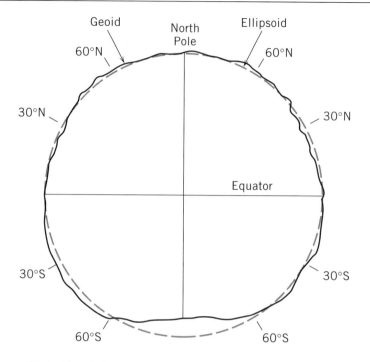

Scale of variations

16. Relation of Geoid to Ellipsoid (After King-Hele, 1963)

GEOID

In the same way that lack of accurate observations delayed recognition of the earth's nonsphericity, so it was not until the advent of artificial satellites and measurements from space that the standard of refinement of collected data would be sufficient to allow scientists to move confidently away from a biaxial figure of reference. Now an even more complicated figure may be the case (figure 16), one with a slight flattening in the northern middle latitudes and a bulge of similar dimensions in the southern middle latitudes. At the time of this discovery, some likened it to a pear shape, but there are those who vehemently oppose the use of such a description, particularly because the deformation is so slight—only 20 m or so—in relation to the overall size.

The term most appropriate to the irregular earth surface is a *geoid*. It does not have a mathematical representation, but it is the *equipotential surface at mean sea level,* which means that at any point it is perpendicular to the direction of gravity. Since that can be affected by variations in the earth structure, the geoid, while a smooth surface, is undulatory. It is not a fully known surface because of the sparseness of gravity observations over large areas of the earth, particularly the oceans. It is the dimension of the equipotential surface at mean sea level that is referred to when quoting the values of the semimajor and semiminor axes.

Many have tried to represent the shape mathematically, but even to get somewhere near requires a series with 18th-power terms. The Defense Mapping Agency developed a mathematical model with terms to the power 180 with 32 755 coefficients. The average (or even the above-average) user can with confidence usually find an alternative way around any problem that calls for manipulating expressions of that magnitude.

In 1993 a new geoid model, Geoid93, was introduced for Puerto Rico, Hawaii, and the Virgin Island. Tests indicated that it was a noticeable improvement on Geoid90. More such figures will no doubt follow.

3.27 cubits = 0.01 stades

$\xleftarrow{\quad d \quad}\rightarrow$

$d = S$ cubits
$= t$ stades

216 Barley corns = 72 inches

$\xleftarrow{\quad d \quad}\rightarrow$

$d = z$ barley corns
$= z/3$ inches

2 yards = 6 ft

$\xleftarrow{\quad d \quad}\rightarrow$

$d = q$ standard yards
$= 3\,q$ feet

0.9387 toise
= 5.632 Paris feet
= 811 lignes

$\xleftarrow{\quad d \quad}\rightarrow$

$d = r$ toise
$= 864.000r$ lignes

1.830 m

$\xleftarrow{\quad d \quad}\rightarrow$

$d = p \times \dfrac{1}{10M}$ quadrant

2842289.6 λ of red cadmium =
3020897.6 λ of krypton-86

$\xleftarrow{\quad d \quad}\rightarrow$

$d = y$ wavelengths

1/163821015 secs =
6.1 × 10^{-9} secs

$\xleftarrow{\quad d \quad}\rightarrow$

$d = x$ seconds

↓

Where
next?

17. Units of Length

CHAPTER
3

Units of Measure

DISTANCE MEASURES

It is one thing to mark points of interest on the earth's surface but quite another to determine how far apart they are. In what sort of unit can the distance be recorded? We are very familiar with the foot as a unit of measure, but in the past numerous lengths of the foot, varying by large amounts, have been recognized in different countries. An even better example of an inconsistent unit is the mile, where a Swedish mile was some seven times the length of an English mile. Thus considerable care has had to be exercised over the centuries to ensure that the particular unit in which a quantity was measured was properly defined. Even so, numerous confusions have arisen.

In earliest times a journey was measured by how long it took to cover the ground; this might be so many camel-days or sailing-days. By the time of Eratosthenes, for example, the distance between Alexandria and Syene was given in terms of the stade. But which particular form of the stade did he use, and what would be its equivalent today? Historians quote different values varying from 157.5 to 210 m, but even these are only approximate equivalents.

Foot and Toise

Both the foot and the toise (the forerunner of the meter) can be traced back to the cubit, (defined as the length of a forearm), some values of which are known to have been in use in 3000 B.C. (figure 17).

Chapter 1 mentioned Arabian miles, Italian miles, and Chinese li, but numerous other units were also used historically. None of these can be given accurate modern equivalents, although attempts have been made to assign unrealistic accuracies to some relationships.

Of the early units the first to be physically represented in any reasonable form were the yard in England and the toise in France.

As early as A.D. 960 the offical standard yard was defined by King Edgar as the length of an hexagonal brass rod kept at Winchester. King Edward I (1239–1307) introduced the first table of relations between measurement units. It started with 3 grains of barley, dry and round = 1 inch and with 3 feet = 1 ulna (represented by a bar of iron made in 1305). The year 1490 saw a new yard standard represented in an octagonal bar introduced by King Henry VII.

In 1588 a ½-inch-square brass rod standard yard was approved by Queen Elizabeth I (1533–1603), based on the previous standards, and this remained standard until 1824.

However, just browsing through papers on historical metrology highlights many different types of foot. The royal foot can be found quoted as equal to 13.75 present inches, the Amsterdam foot as 11.41 inches and the natural foot as 9.8 inches. The time of the building of Stonehenge is probably as far back as we can go with any degree of certainty. Sir Flinders Petrie, investigating in 1880, found the inner diameter of the circle of faces of the sarcen stones to be 1167.9 inches, which is almost exactly 100 times the Roman foot of 11.67 inches. This foot was descended from the royal cubit of 20.63 inches.

The French, not to be outdone, had a foot measure descended from the Assyrian cubit of about 21.6 inches. This cubit later became the Hashimi cubit, a half measure said to have been introduced into France by Charlemagne in around A.D. 790. It was then known as the pied du Roi and survived for 1000 years.

The pied du Roi of 12 pouces equates to 12.79 present inches. Six pied du Roi became the toise of 6.395 British feet, a unit that was later used to measure both the northern (Lapland) and southern (Peru) arcs and other arcs of the 17th and 18th centuries. Its most notable physical representation prior to the expeditions was as an iron rod set in the stairs of the Châtelet in Paris.

From the middle of the 17th century there was a growing realization of the need for a universal unit of measurement. The problem was how to define such a unit. It needed to be in a form that could be readily reproduced, but at that time there was no obvious candidate.

A pioneer in this field was Gabriel Mouton, a vicar from Lyons, who in 1670 proposed that a minute of arc of a great circle be adopted as a universal unit. Mouton, like other prominent scientists of those times, was a Jesuit priest who treated astronomy as a hobby. He used the work of another Jesuit, John Riccioli, to illustrate his idea. In essence it consisted of the following relations:

1 milliar	=	1000 virga
1 centuria	=	100 virga
1 decuria	=	10 virga
1 virga	=	0.001 minute of arc
1 decima	=	0.1 virga
1 centesima	=	0.01 virga
1 millesima	=	0.001 virga

The decima was to have the same length as a pendulum that made 3959.2 oscillations in 30 minutes at Lyons, but nothing came of these efforts.

In England Sir Christopher Wren had similar aspirations for a universal unit and was considering the length of a half-second pendulum in London as a possibly appropriate unit.

Over the English Channel, the toise du Châtelet had become some 5 lignes (where 1 *ligne* = ¹⁄₁₂ inch) too long through usage and wear. In 1671 l'Abbé Picard, while occupied in restoring the toise, worked along the same lines and suggested that a universal foot should be ⅓ of the length of a second pendulum (see Chapter 11) or "rayon astronomique" at the Paris Observatory. This was 36 pouces, 8.5 lignes in length where the Paris pouce (inch) was equivalent to about 1.065 present inches. A similar pendulum suggestion came in 1673 from the Dutch astronomer Christiaan Huygens.

The scent went cold for some 50 years until Jacques Cassini II in 1720 reverted to the notion of using the length of a minute of arc by suggesting that a geodetic foot be 1/6000 of a terrestrial minute of arc. This would have been very close to the approximation used in some present-day calculations of 1″ (second of arc) = 100 ft. Meanwhile the toise was reconstructed specifically for the expeditions of 1735 to Peru and Lapland as the toise du Perou and the toise du Nord, respectively. They were in the form of a flat iron bar of 864.000 Paris lignes (where 1 pied du Roi = 144 lignes) or 6 Paris feet.

Meter

As a result of the Peru expedition, La Condamine recommended in 1747 that the equatorial seconds pendulum should be used to define a standard. This he found to be 439.15 lignes. He was emphatic in saying that it should be an international unit and that perhaps a group of academics from various countries should investigate the problem.

In 1789 a conference of delegates from numerous French towns petitioned Louis XVI to unify the system of measures. The year 1790 saw the proposal of a new unit to be called the *metre*, based, yet again, on the seconds pendulum.

The impetus for a start came from Prieur Du Vernois, later known as

C.-A. Prieur de la Côte-d'Or. On February 9, 1790, he presented to the National Academy a memoir on the necessity for a uniform set of measures. This occurred during the difficult times of the French Revolution, but he was adept at the politics necessary to promote his ideas without risking the guillotine. He proposed relating the unit to the length of the seconds pendulum at the Paris Observatory. A platinum bar deposited at the City Hall would represent the length. One-third of its length would represent the national foot, which in turn would divide into 10 inches, each of 10 lignes. Ten feet would equal 1 rod.

Charles Maurice de Tallyrand, bishop of Autun and later president of the National Assembly, put Prieur's ideas to the Assembly. In doing so he suggested using the length of a seconds pendulum at 45° latitude so as to be independent of any particular location.

Around the same time, both the United States and Great Britain were investigating the possibilities of a new system of measures. As a result it was suggested that the British and French parliaments, the French Academy of Sciences, and the British Royal Society should cooperate in creating the new system, but this did not materialize.

The members of the committee of the French Academy of Sciences including such well-known names as Jean Borda, Pierre Laplace, Joseph Lagrange, Gaspard Monge, and Marie de Condorcet. Their first report in March 1791 particularly emphasized that on no account should any existing unit of measure be used as the new standard since it would only generate national pride in the lucky country but hurt in every other country.

Among the methods they considered were:

1. a seconds pendulum at the equator
2. a fraction of the circumference of the equator
3. a fraction of a meridian quadrant

They rejected the first because of the influence on it of time and gravity variations and the second because of the difficulty in measuring along the equator. That left only the meridian arc. It was decided that this would be derived from a new meridian measured from Dunkirk to Barcelona that would pass through the Paris Observatory.

The National Assembly approved the Academy's plan in March 1791 and as a result the Academy established five commissions to investigate triangulation, observation of latitude, baseline measurement, length of the seconds pendulum, comparison of units, and standardization of weights.

On the eve of his flight from Paris, the king discussed the projects with the commissioners and in particular with Cassini, whose father and grandfather had previously measured a chain of triangulation along the meridian through Paris. Cassini pointed out that the instruments available to these new efforts were much improved over those used by his ancestors.

Field Measurements

A year was spent on the production of suitable equipment, among which was the celebrated Borda circle and other equipment constructed to his design. By the end of 1792 Jean Delambre had started on the section from Dunkirk south to Rodez, and Pierre Mechain was beginning in the Pyrenees, an area he had not previously visited.

As a war was still in progress, life was not easy for the surveyors. They were looked upon with suspicion by villagers and on many occasions were arrested as spies. Measurements for the 9°39′ arc took until 1798 to complete. While the survey work was still in progress, the Third Decimal Committee was anxious to establish a provisional unit. A report from the Academy to the convention in May 1793 recommended dividing the new units decimally, giving the units simple and internationally intelligible names, and assigning the meter the provisional length of 443.443 Paris lignes. This last was based on l'Abbé Nicholas La Caille's value of 57 027 toises for 1° of the meridian at 45° latitude. A brass standard of this length was produced by Jean Borda and Mathurin Brisson.

However, when the arc was finally completed, the commission gave the relation as 1 meter = 443.295 936 Paris lignes at 0°C, and a new scale was ordered from Lenoir. This gave 1 French foot = 0.324 839 meter.

The precision with which Delambre and Mechain measured their arc was about 5×10^{-4}. The metre des Archives that resulted from the arc measures was made in 1799 to a precision of about 5×10^{-5}, or about one order higher than the actual arcs. It was around 1850 that the precision of calibrations improved by another order of magnitude, and the 1880s before 10^{-7} was reached.

Definitive Value

When the results became available the assumption was made that the earth was an exact spheroid and that the ellipticity, dimensions, and quadrant length should be determined from this new arc together with that of Bouguer and La Condamine in Peru. The result gave the ellipticity as 1/304 and the meter as 443.296 lignes.

The data they used to achieve these values were:

	Arc length (toises)	Amplitude	Mid. lat.
for France:	551 584.7	9°40′25″	46°11′58″
for Peru:	176 873	3 07 01	1 31 00

The approximate solution can be found thus:

1. If $2A$ and $2B$ are the sum and difference of the semimajor and -minor axes, then neglecting powers of B/A we get

$$R = A - 3B \cos 2\phi$$

2. Integrating from 0 to $\pi/2$ gives the length of the quadrant:

$$Q = A\pi/2$$

3. From the radii at the two latitudes ϕ_1 and ϕ_2 there are two equations from which B can be eliminated. Then

$$A = \varepsilon + \omega \cot \sigma \cot \delta$$

where 2ε and 2ω are the sum and difference of the radii, and σ and δ are the sum and difference of the mean latitudes.

4. From these data the curvatures of centers of the two arcs are

$$3\ 266\ 977 \quad \text{and} \quad 3\ 251\ 285$$

Whence $\varepsilon = 3\ 259\ 131$ and $\omega = 7846$.

5. From this we get

$$A = 3\ 266\ 345$$

and multiplying by $\pi/2$ and dividing by 10^7 gives the meter as 0.513 0766 parts of the toise du Perou (figure 17).

6. Since the toise $= 864$ lignes, the meter became 443.296 lignes.

The Legal metre

The metric system was legalized in France on April 7, 1795, and resulted in the expression "Legal metre." Remember however, that the meter was at first defined only as a certain proportion of the toise du Perou and did not actually materialize until Borda's metre des Archives. At 0°C this defined value was 443.296 lignes of the toise du Perou when the latter was at 16.25°C; that is, 443.296/864 became the old French legal meter. This gave the relation

$$1 \text{ toise} = 1.949\ 0366 \text{ m} \qquad \text{or} \qquad 1 \text{ m} = 0.513\ 0740 \text{ toise}$$

Since at the outset the meter was not represented by any bar, the Toise du Perou remained the national material standard. That is, initially the meter was simply defined by law, so when countries wanted copies, they were issued copies of the Toise du Perou.

In 1798 Borda made the metre des Archives, which was a flat bar of impure platinum in the form of an end standard. When it was found that the length of the meter was impossible to redetermine accurately in terms of its physical basis, the metre des Archives became the basic standard. Actually this meter was one of three platinum standards and several iron ones made at this time.

At about the same time the French Academy considered the toise du Perou unsatisfactory, so it instructed Borda to prepare new bars. He made four bimetallic rules of strips of copper and platinum. The length was a double toise of about 12.8 English feet, No I of which became known as the module and was regarded as replacing the toise du Perou.

On June 22, 1799, the new meter standards were formally presented to the Corps Legislatif and legalized by a statute that abolished the provisional standards. Hence from about 1800 there were two standards in existence in France—the toise and the meter.

In the 1810s there was much mistrust of the new units of measure. Napoleon found the popular disregard for the new system so great that he sanctioned a parallel system that applied the old names and fractions to the new units. In 1812 there came a French imperial decree "for facilitating and accelerating the establishment of a uniform system of weights and measures in our Empire." The decimal division was abandoned in favor of binary and duodecimal divisions of metric units.

Soon, however, the decree of 1812 was repealed, and on July 4, 1837, the use of any units other than decimal ones became an offense with effect from January 1, 1840. In 1841 Friedrich Bessel gave the earth quadrant as 10 000 856 m, so that already the indefiniteness of the standard was becoming evident. By 1857 Alexander Clarke was quoting a quadrant of 10 001 983 m.

Despite the introduction of the meter, standard yards and other units persisted.

In 1864 a committee was formed from delegates at an international exhibition in Paris to bring uniformity into measures. This goal was further pursued at the Geodetic Association meeting in Berlin.

In 1869 the French government invited foreign governments to attend a conference in Paris on length standardization. The year 1870 saw the first meeting of the International Metric Commission, and in August of that year it was resolved to establish the Bureau Internationale des Poids et Mésures

(BIPM) and to construct a number of precise line standards as similar as possible to the metre des Archives. From here the meter gradually began to gain international acceptance, although, as we all know, there are still those who are not happy using either it or its relatives.

Various intercomparisons of standards and manufacture of copies took place and in 1889; 30 copies of the prototype meter were distributed to different countries. For example, numbers 11 and 28 went to Russia, number 16 to the United Kingdom, and numbers 21 and 27 to the United States.

A Natural Standard

By 1890 Sir David Gill was agitating for a natural standard and suggested the wavelength of sodium vapor as a possibility. There were other suggestions before Albert Michelson and Justin Benoit in 1893 defined the meter in terms of a specific number of wavelengths of red cadmium radiation (figure 17).

A comparison of the meter with the yard in 1894–1895 remained the U.K. standard for the meter until 1964.

In 1946 the Empire Scientific Conference redefined the yard as exactly 0.9144 m. Prior to this time the U.S. inch was 25.400 05 mm and the British inch was 25.399 93 mm.

In 1960 the meter was redefined in terms of a wavelength of light as 1650 763.73 λ of krypton-86 gas in a vacuum (figure 17). A line in the orange part of the visible spectrum corresponding to the unperturbed transition between $2P_{10}$ and $5N_5$ levels. This allowed a comparison of standard lengths that was accurate to 4 parts in 10^9.

In 1983 the General Conference on Weights and Measures meeting in Paris redefined the meter yet again, this time in terms of a time standard. It is now the length of the path traveled by light in a vacuum during a time interval of 1/299 792 458 of a second. Notice that this number is the reciprocal of the accepted value of the velocity of light. Comparison is now possible to better than 1 part in 10^{10}.

Thus the 20th century has seen several shifts in defining units. In the 1940s accurately measured distances were used to determine the velocity of light. Later this was turned round so that the velocity could be used to determine distance. We now have distance defined in terms of time.

ANGLE MEASURE

In parallel with the introduction of the meter came the *gon*, a decimalized angular unit. A right angle contains 100 gons; hence the full circle is 400^g. Instead of the *sexagesimal* (60) minutes and seconds, a gon subdivides into 100^c (*centesimal* minutes), each of 100^{cc} (centesimal seconds). Thus an angle will be written as 379.4582^g or $379^g\ 45^c\ 82^{cc}$. Note that most modern calculators incorrectly call this unit a "grad" or "grade."

Introduction of the gon does not mean that the sexagesimal degree system has gone. It is still used in parallel with the gon system. When you think about it, sexagesimal divisions affect not only angles but also time, latitude, and longitude, so any move to sole use of the gon would have enormous repercussions in measurements.

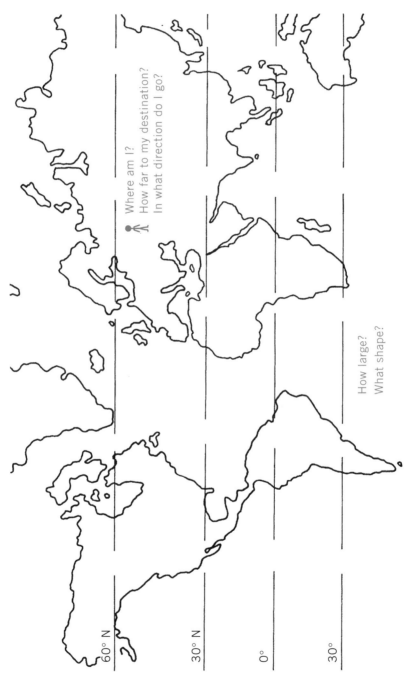

Where am I?
How far to my destination?
In what direction do I go?

How large?
What shape?

18. Questions about the Earth

CHAPTER
4
Traditional Survey Positioning Techniques

Once knowledge is available on the size and shape of the earth, the next question is "Where are we on that earth?" or "Where on earth are we?" (figure 18).

One of the main functions of geodesy is to determine the exact positions of points on or near the earth's surface. This includes points below the surface for mining or tunneling, under the oceans for underwater engineering structures or oil wells, and above the surface for rocketry or radio telescope positioning.

Recent years have seen the traditional techniques supplemented by new ones. It would now seem to be only a matter of time before the older methods are fully superseded by satellite methods and the need for permanent survey marks disappears. This is not to suggest that all marks would be destroyed or let fall into disrepair since undoubtedly there would still be small operators who would require them.

This chapter will concentrate on the time-honored approaches, and later chapters will cover the newer techniques. While there is interest in all three spatial dimensions, it is convenient to separate the horizontal positioning from the vertical element since in traditional survey work they tended to be separate operations. With the newer methods this is not so since the data used can normally solve for all three dimensions at the same time.

Horizontal positioning methods include

1. astronomical techniques   3. trilateration
2. triangulation   4. traversing

while positioning in the vertical plane includes

1. precise or geodetic leveling,   3. barometric leveling
2. trigonometric leveling   4. echo sounding

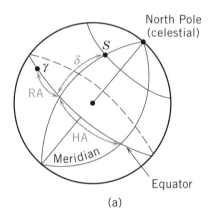

(a)

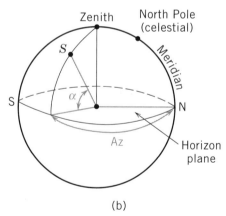

(b)

19. Celestial Coordinate Systems

HORIZONTAL POSITIONING

Astronomical Techniques

From the very early days of civilization the heavens have been used to aid positioning. Until the last few centuries the N-S component, or *latitude*, was far easier to find than was the E-W component, or *longitude*. While the latter had to await the invention of good chronometers getting an indication of latitude was relatively straightforward using simple stellar observation. Initially astronomy was used for marine navigation; later explorers found it very useful in uncharted regions. Then the geodesist used it in combination with other measurements to establish precise positions on which to base national surveys.

Single isolated fixes by astronomical methods were of little value to the surveyor, because they could not be interrelated. But taken at positions that were connected by other means, they were of extreme importance.

From early times it was recognized that the stars appeared to rotate around the heavens and that there was a star near to the apparent hub of that rotation. Even before the idea of an Equator was fully appreciated, it was observed that by traveling north or south the elevation of this hub star (Pole Star) changed by amounts that could be related to the distance traveled. Unfortunately for the surveyor, as well as the navigator and explorer, there is no equivalent star visible in the Southern Hemisphere that gives a direct approximation to the position of the South Pole. The nearest star there is σ Octantis in the constellation known as the Southern Cross, but its relationship to the pole at any time requires computation.

In the early centuries after Christ, strenuous efforts were made to catalogue as many of the heavenly bodies as possible in terms of their relative positions. Primitive observatories came and went as the importance of detailed knowledge of the movements and positions of the stars and planets came to be appreciated.

Among the methods used for describing the positions of the stars on a celestial sphere are those shown in figure 19:

1. by right ascension (RA) and declination (δ) (figure 19a)
2. by hour angle (HA) and declination (δ) (figure 19a)
3. by azimuth (Az) and altitude (α) (figure 19b)

where

right ascension = the angle along the celestial equator from an initial point called the first point of Aries (γ) to the declination circle of the body

declination = the arc between the body and the celestial equator

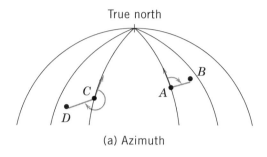

(a) Azimuth

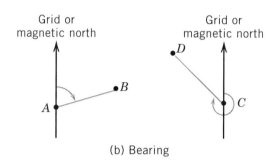

(b) Bearing

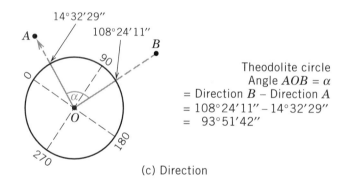

Theodolite circle
Angle $AOB = \alpha$
= Direction B − Direction A
= 108°24′11″ − 14°32′29″
= 93°51′42″

(c) Direction

20. Azimuth, Bearing, and Direction

hour angle	= the angle between the plane of the meridian through Greenwich and the plane through the body
azimuth	= the horizontal angle along the horizon plane between the meridian of the observer and that through the body
altitude	= the vertical angle from the horizon to the plane of the body

Of the three systems, the first is used in tabulating star catalogues since it is independent of the observer's position. The other two systems are used for field observations in conjunction with determining point position and orientation.

The 17th century saw the formation of important scientific bodies, such as the Academy of Sciences in France and the Royal Society in London, to encourage these among numerous other aspects of science. It also saw the foundation of the Royal Greenwich Observatory in 1675 under warrant from King Charles II and with John Flamsteed as "astronomical observator," and other major observatories on the continent of Europe. These were great stimuli to production of accurate tables of the stars (ephemerides) and to improvements in the methods of recording time.

Astronomical methods for positioning yield results essentially in terms of latitude and longitude, although they may at a later stage be converted into a rectangular coordinate system of some form. With these two elements we are also normally interested in the azimuth of a particular line, which can be found from similar observations.

Azimuth, bearing, and direction   Three terms need to be distinguished (figure 20). The *azimuth* from point *A* to point *B* is the clockwise angle measured at *A* between the plane containing the North Pole and the plane containing *B*. There are subtle differences between astronomical azimuth and geodetic azimuth. The former relates to the geoid, while the latter relates to the adopted ellipsoid.

The term *bearing* is normally retained to indicate the clockwise angle from grid north when working in terms of rectangular coordinates.

Direction is used to denote the circle reading on an instrument, such as a theodolite, and the difference between two recorded directions gives the angle at that point.

We must accept, however, that, while it is preferable for all nations to use the same definitions and terminology, it is a fact of life that neither these nor some other survey terms are universally exclusive, and so should be interpreted in the context of the publication in which they appear.

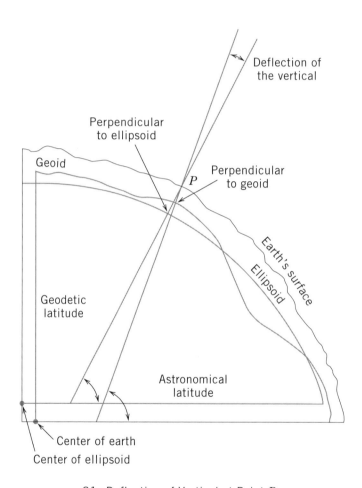

21. Deflection of Vertical at Point P

True north, grid north, and magnetic north The previous section made mention of grid north. Surveying uses several different "norths," so the differences need to be known.

True north is the direction from a point toward the north end of the axis of rotation of the earth, and is nearly equal to the direction of the Pole Star or North Pole.

Grid north is the north direction of a set of rectangular coordinates and may (though more likely will not) be coincident with true north. It can be a quite arbitrary direction.

Magnetic north is the direction indicated by a magnetic compass. It is not constant, and if shown on maps is usually accompanied by an indication of its relation to grid north and the amount by which it is changing annually.

Latitude There are many ways the surveyor can determine his latitude through astronomical observations aside from that already indicated, where the altitude of the Pole Star closely approximates the required result.

It follows from this that if the *altitude* (the vertical angle from the horizon to the body) is measured to a star or the sun and we know the position of the body (because it has been catalogued) in relation to the celestial pole, then we can determine the latitude. For the best results the observations require correction for refraction. If it is the sun being observed, then allowance is needed for its *semidiameter;* that is, when viewed through a theodolite telescope, the sun has a finite diameter that fills a large percentage of the field of view, whereas a star can be considered as a point.

Among the alternative methods is the use of *circummeridian* observations. A series of readings are taken to either the sun or a star over a period from some 10 minutes before until 10 minutes after it has crossed the meridian. With accurate knowledge of the time of each observation, corrections can be applied to get the equivalent values at the meridian. With suitable observing techniques, a high degree of precision is possible.

Longitude For centuries longitude proved to be a much more difficult nut to crack. Early explorers and navigators made errors of many degrees in their longitude position because of the methods they had to adopt. Longitude is essentially a function of time—a quantity that could not be measured sufficiently accurately until the second half of the 18th century. Although the ancient Chinese had their water clocks and other cultures their sand glasses, these were not timepieces that could be transported from place to place. Positioning required knowing the difference in time between a datum position—nowadays this is the Greenwich meridian, but in the past many other positions have been used, including Paris and a point in the Canary Isles—and the point in question.

Before the introduction of radio signals, the only way of getting the time from *A* to *B* was to physically carry it in the form of a timepiece. While early clocks had reasonable accuracies when stationary, they were completely upset

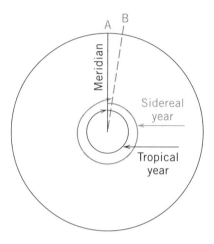

(a) Tropical and Sidereal Year Defined

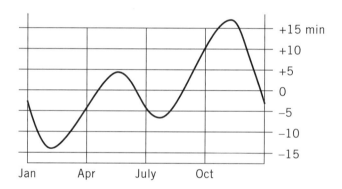

(b) Difference between Apparent Solar Time and Mean Solar Time

22. Different Concepts of Year

when taken on board ship. So much so that a reward was offered by the government for a solution to the longitude problem—in effect for building a timepiece that would keep accurate time on a long sea voyage. Other solutions were suggested, but were not at all practicable.

It was John Harrison who produced the first acceptable chronometer in 1765 and who, after a considerable struggle, managed to claim the reward. Nearly another 100 years were to elapse (1852) before it was possible to send time signals by electric telegraph, and it wasn't until the turn of this century that wireless telegraphy was introduced. Nowadays it is possible to receive radio time signals throughout the world to a very high degree of accuracy.

When considering time and longitude it must be remembered that 1 second of time is 15 seconds of arc, or around 0.5 km at the equator. The difference in longitude between two places is then the difference between the times at those places at the same instant. This is the same as saying the difference in time between a particular star being directly overhead at the Greenwich meridian and the same star directly overhead at the point.

By using star or sun observations the local time at a particular point can be found, and the requirement is then to find the local time at the reference station at the same instant. This can be done using either time signals or a portable chronometer.

Azimuth can similarly be found by several astronomical methods. Since the stars appear to circle with the Pole Star as a hub (almost), then by noting the positions of a particular star when it is farthest east and farthest west—in other words, at elongation—we can find the meridian direction, which is clearly halfway in between. Next we record the angle between the star and a ground point. Finally, we find the azimuth of the point of observation by noting the angle between that point and the ground point.

Astronomical observations made with a theodolite or similar instrument all relate to some leveling device on the instrument. Thus, when properly adjusted, the vertical axis of the instrument coincides with the direction of gravity and is therefore perpendicular to the geoid (figure 21). Thus astronomical positions are referenced to the geoid. Since the geoid is an irregular, nonmathematical surface, astronomical positions are wholly independent of each other. If sufficient knowledge of the geoid and ellipsoid is available, then conversions are possible.

Time   Modern requirements for time have far outstripped the accuracy that can be obtained from a pendulum clock. As the explosion in technology has gathered pace, so better and better timepieces have become necessary to achieve finer and finer resolutions. Time is of course of particular interest because it is also a measure of longitude: since $360° = 24$ hours, $1° = 4$ minutes.

Year   Since the sun returns to a different position after a year, there are two different concepts of year. The *tropical year* is the time required for the sun to return to the same position as the previous year in relation to the *equinoctial points* (the points where the equator intersects the ecliptic). And the *sidereal year* is the time it takes the sun to return to the same position as before in relation to the stars. In terms of figure 22a, this means that the sun goes from point A to A for the tropical year from A to B for the sidereal year. Thus,

$$\text{sidereal year} = 365.242\ 199 \text{ mean solar days}$$

$$\text{tropical year} = 365.256\ 360 \text{ mean solar days}$$

a difference of about 20 minutes. Where a *mean solar day* is the interval between two successive transits of the mean sun over a given meridian, and the *mean sun* is a fictitious body, describing a circuit of the equator in the time required for the actual sun to describe the ecliptic. The difference between apparent solar and mean solar time at any instant is given by the equation of time, which varies between -14.25 to $+16.25$ minutes (figure 22b).

This all begins to get confused with different types of sun, day, and year but these stem from the nonuniform motion of the actual sun in its apparent orbit. It is much easier to deal with a hypothetical sun moving at the average velocity of the actual sun—in other words, with the mean sun.

The big problem with time is that there is such a bewildering range of forms in which it can be expressed. When writing this text, I had no difficulty amassing a list of 25 different forms of time that could warrant mention. To describe them all in detail here would have little effect other than to create confusion. What it does highlight is how careful we have to be to know which system we are working in and how to convert from one system to another if necessary.

Day   The time unit day is defined as the interval between successive transits of a celestial body across the same part of a meridian. There are three versions depending whether the body is the sun (solar), a star (sidereal), or the moon (lunar). All are slightly variable. The last can be ignored here.

Solar time   This is the natural clock in terms of the daily motion of the sun. The *solar day* is the interval between successive upper transits of the sun, but for various reasons this is not a constant. Using the mean solar time (a function of the average of all apparent solar days throughout a year) overcomes this irregularity. This gives rise to an imaginary mean sun moving at a uniform rate round the equator.

Sidereal time While the sidereal time is the hour angle of the first point of Aries, it is also equal to the hour angle plus the right ascension of a given star.

Greenwich time As the prime meridian is taken to be that through Greenwich, England, so various measures of time are called Greenwich time. This may be sidereal, apparent solar, or mean solar time. The basic relation is that

Greenwich time = local time + longitude of the place of observation.

This leads to the concept *standard time,* which is the time adopted for different bands of longitude for everyday purposes. Whence

standard time = Greenwich mean time ± an integral number of hours

Greenwich mean time (GMT) is now often referred to as *universal time* (UT). Although there is still an observatory at Greenwich, its time, denoted UT 1, is now the mean value derived from some 54 observatories around the world. A modification of this, to allow for annual variations in the earth's rotation, gives UT 2, where the difference between UT 1 and UT 2 varies by a few hundredths of a second of time over a year.

$$\frac{\text{Greenwich apparent sidereal}}{\text{time (GAST)}} = \frac{\text{local apparent sidereal}}{\text{time (LAST)} - \lambda}$$

where λ is the longitude of the point of observation.
GAST can be converted to universal time, as UT 0, by using a set of tables. Then

UT 1 = UT 0 + a small reduction to the conventional international origin (CIO)

UT 2 = UT 1 + a small correction related to the polar motion

It is UT 1 that the surveyor or geodesist uses when reducing his astronomical observations.

Ephemeris time To accommodate the higher accuracies required by dynamical astronomy, geodesists replaced mean sidereal time with ephemeris time and atomic time (AT).
Ephemeris time (ET) is related to UT 1 via a small correction found in the Astronomical Ephemeris, where

$$ET = UT\ 1 + dT$$

The correcting term was zero in 1902 and +34 seconds in 1961, but it does not increase uniformly.

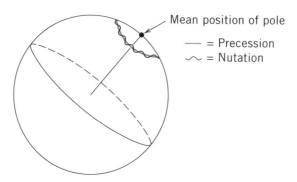

Mean position of pole
—— = Precession
⌇ = Nutation

(a) Precession and Nutation

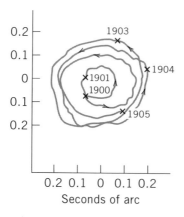

(b) Polar Motion (1900–1905)

23. Motions of Earth's Axis

Atomic time Modern atomic clocks are far more stable than any previous form of timepiece. Because of this, they are able to detect irregularities in the earth's rotation. Hence the introduction in 1967 of the concept *atomic time* (AT, now sometimes called the SI second), which began as the weighted mean of eight atomic standards.

It was defined that year such that 1 second = 9 192 631 770 cycles of a caesium-133 atom under defined conditions. The atomic second is very close to an ephemeris second. *International atomic time* (TAI) was arranged to have a distinct relation to ephemeris time as January 1, 1977, and has since replaced the latter.

Coordinated universal time (UTC) is directly related to atomic time with a periodic jump of 1 second to keep it within that amount of UT 2.

Other forms of time As you investigate other relevant publications it will become obvious that there are yet more versions of time. It is not appropriate to refer to them all here, but you must follow them if you're aiming to become a specialist in geodesy.

Time in context of global positioning system (GPS) Each satellite contains several atomic frequency standards since the highest-possible orders of accuracy are required. These are atomic clocks based on both caesium and rubidium. The latter is the cheaper of the two versions. Despite their high accuracy, they are still subject to a steady drift away from the GPS standard. This variation can be closely tracked at the master control station in the United States and its value incorporated in the broadcast signals. The magnitude of such drift is kept to within 0.001 second.

The ground receivers normally do not have atomic frequency standards but instead use quartz crystal clocks. Nevertheless, the stability is of a high order—better than 1 part in 10^{10}, which means better than 1 second in over 300 years. Today this can be bettered to even 1 in 10^{15}, an accuracy almost beyond comprehension.

Remember also that since the velocity of light is 300 000 km/second, 10^{-6} second (= 1 microsec) is equivalent to 300 m. Hence, to get down to accuracies of centimeters, resolutions to better than 10^{-10} seconds are essential. While it is not possible to sychronize the various clocks to this accuracy, there are ways to reduce the effect of such synchronization errors.

Terms

Those small troublesome terms The earth's axis is, unfortunately, not fixed in space, and neither are the positions of the apparently "fixed" stars. These effects result particularly from the fact that the earth is not a true sphere. As a result, small correcting terms are required, each amounting to a few seconds of arc. To complicate the issue still further, the corrections are not constant but vary with time. A brief description of each appropriate one follows, although their effect will often come built into any software package the geodesist uses.

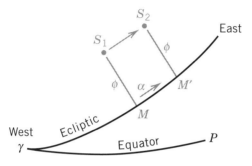

(a) Apparent Change: M to M'

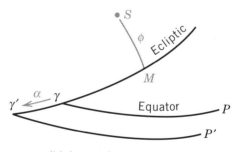

(b) Actual Changes: γ to γ'

24. Precession

Aberration James Bradley spent a year starting in December 1725 observing a particular star through a fixed telescope. He found that it seemed to move with time in such a way as to complete a cycle in a year and moved as much as 20″ of arc from its original position. He tried various hypotheses including nutation (see below) to account for this change, but to no avail until 1728. It was then, after further observations of more stars, that he hit upon the idea that the effect might result from the finite velocity of light combined with the motion of the earth in its orbit. The apparent position of a star is determined by the movement of the light from that star relative to the earth, which will vary from star to star. Any one star traces out an approximate ellipse around the point where it would be seen if there were no *aberration*. Over the years various estimates have been put on the maximum value of aberration, all of which fall in the vicinity of 20.4″. Working backward from his results, Bradley concluded that the velocity of light exceeded the velocity of the earth in its orbit by a factor of 10 210 to 1 and hence that light took approximately 500 seconds to travel from the sun to the earth.

Since the velocity of the earth was then known to be about 18.3 miles/sec, Bradley's values would suggest a velocity of light of 18.3 × 10 210 = 186 850 miles per hour—not a bad estimate compared with today's knowledge.

Nutation In addition to the complications caused by the continual variation of the separation of the attractive bodies (precession, see next section), there is a small oscillation of the earth's axis, called *nutation*. This also was discovered by Bradley from observations made over the 20-year period of 1727–1747. He found an irregularity in motion of the pole that had a period of 18.6 years and which became known as nutation. The pole was describing an ellipse with axes of 18″ and 16″ (today taken as 18.42″ and 13.75″), which gave the precessional movement a superimposed wavy or periodic variation.

Precession Compared with observations made 150 years before his time, Hipparchus (c. 150 B.C.) found there was a small general increase in the longitude of stars but no changes in their latitude. He estimated this change to be about 36″ a year (from west to east). From this he deduced that the positions of the equinoctial points were changing and that it must be the equator that was slowly moving.

In this situation, the sun will return to a slightly different point at the completion of each year, a position a little before that of the previous year (figures 23a and 24).

In fact, the attractive effects between the earth and other heavenly bodies, particularly the sun and moon, cause this slow movement, known as *precession of the equinoxes,* and which amounts to about 50.2″ a year (actually 50.2564 + a small term). Hence over the not inconsiderable period of 25 800 years [(360° × 3600)/50.2] the celestial pole will describe a circular motion around the pole of the ecliptic. The motion can be likened to a spinning top: while it rotates on a point, its uppermost part also rotates to sweep out a conical shape.

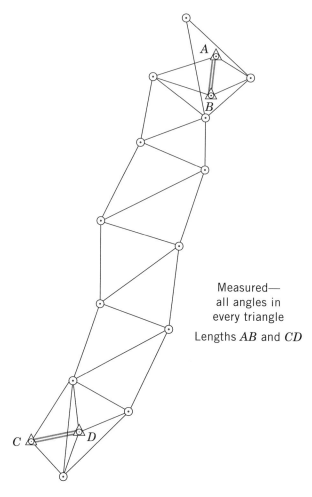

Measured—
all angles in
every triangle

Lengths *AB* and *CD*

25. Triangulation Network

While Hipparchus was the first to conclude that such a change occurred, it was left to Newton to come up with the right explanation: because the sun seldom lies in the plane of the earth's equator, most of the time its attractive force is not directed toward the center of the earth.

Polar motion   Observations in Berlin 1884–1885 indicated minute variations in latitudes of about 0.5″. Dr. S. C. Chandler showed in 1891 that they could be explained if the earth's axis described a circuit around its mean position in about 434 days. The displacement was never more than 10 m and has not yet been sensibly explained, but it is known as the *Chandler period*. It would appear that the earth's axis of rotation is moving by a very small amount in relation to the earth's crust. Figure 23b shows an example over a longer period.

Aberration, nutation, precession, and polar motion—these are just a few of the sometimes mystifying corrections that must be made either in astronomical observations to stars or in observations involving satellites.

Triangulation

Since the 1620s, when Snellius developed the idea of triangulation (figures 7 and 25), it has been used extensively throughout the world. It has formed the skeleton upon which most national surveys have been based. Triangulation links together a series of triangles and quadrilaterals formed between a number of pertinent points over the area to be mapped. These points, because of requirements of the observational methods, tend to be prominent positions such as pillars on hilltops, church towers, and similar high points.

Essentially, as the name implies, *triangulation* requires the measurement of three angles in each triangle. A little more than that is needed, however, to get scale, position, and orientation into a scheme. So that everything is correct to scale, at least one accurately measured line is needed. In schemes of any size there are often two, one towards each end of the chain of triangles as a check on the accumulation of inaccuracies. Positioning the whole chain correctly on the surface of the earth requires finding the coordinates (latitude, longitude, and height, or *easting, northing*, and *height*) of at least one point and the azimuth of at least one line.

In a national survey scheme, the triangle sides are likely to be on average 50 km (30 miles) long, but they have been known to be over 100 km. On the other hand, any baseline would be about 15 to 20 km (10 to 12 miles), because of the difficulty in finding suitable sites and carrying out the actual measuring procedure. The baseline would, however, be measured with every conceivable precaution.

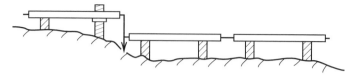

(a) 18th and 19th Century — by wooden bars

(b) First Half 20th Century — by invar wire in catenary

(c) Today — by electromagnetic distance measuring

26. Equipment for Distance Measurement

Baselines The baselines of triangulation nets observed in the 17th to 19th centuries were measured with a wide range of different equipment (figure 26): wooden rods placed end to end, glass rods, metal bars, steel chain, bimetallic rods, and variations on these. One problem with all these methods was accounting for the expansion of the materials with changes in temperature.

Toward the end of the 19th century came a major breakthrough with the invention of *invar*. This is a nickel-steel alloy whose low coefficient of expansion is some 10 times less than that of steel. This material could be used in the form of long wires (up to 300 ft) suspended between tripods as a catenary and under known tension conditions. Accuracies of the order of 1 in 1 million (1 mm in 1 km, or 1 inch in 15.8 miles) proved possible, so the method was used to measure all important baselines up to the 1960s.

Angles The angles in triangulation were measured with the best available theodolites—often reading to 0.1″. This allowed computation of the lengths of every other side in the scheme by simple trigonometry.

From the initial orientation (probably from astronomical observations) the orientations of every line could be calculated. By combining these data with the given initial coordinates, the coordinates of every point could be computed. Various corrections and adjustments were applied to make whole scheme as mathematically sound as possible. After all, so much future work would depend on this skeleton that it was wise to take all feasible care and effort.

Major early triangulation networks covered India, Great Britain, and large parts of Europe. The United States and large parts of Africa followed. A modern map would show most of the world covered either by triangulation or the following alternatives of trilateration and traverse.

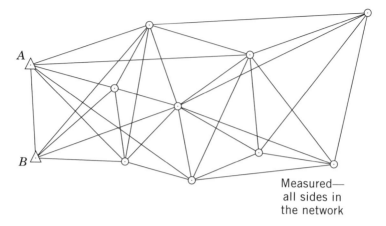

Measured—
all sides in
the network

27. A Trilateration Network

Trilateration

As the name implies, trilateration requires the measurement of three sides in each triangle. The idea is the same as for triangulation in that it is a method of interconnecting, and hence positioning, a number of points in relation to one another. With the advent of electromagnetic methods for measuring distance it became quicker and more economic to record distances than angles (figure 27).

The basic data required is either the coordinates of A and B or the coordinates of one point and the azimuth to the other. In practice, a scheme is likely to involve a mixture of both triangulation and trilateration (sometimes referred to as *triangulateration*) where a majority of the sides are measured together with selected angles. The appropriate mix can often be determined through a network optimization exercise—or computer simulation—of different configurations.

The *electromagnetic distance measuring* (EDM) equipment that has made trilateration feasible did not come into wide use until the 1960s. It was first developed in the late 1940s as a result of experiments to determine the velocity of light. Such experiments required accurately measured distances against which to test the observations. Later, when the velocity became determinable to an accuracy equivalent to that of the measured distance, the whole idea was turned around. Particularly under the initiative of Erik Bergstrand in Sweden, *the Geodimeter* was developed, and the first production model appeared in 1949. It was extremely heavy and bulky, but those national survey departments around the world that tried it out achieved very encouraging results.

By 1957 Dr. Trevor Wadley in South Africa had developed a similar instrument—the *Tellurometer*—that operated on microwaves. In the beginning, the observational and reduction procedures were tedious and time-consuming, but as electronics developed, the equipment became more readily acceptable.

As with triangulation, so trilateration requires at least one set of known coordinates and an orientation on which to found the computed coordinates.

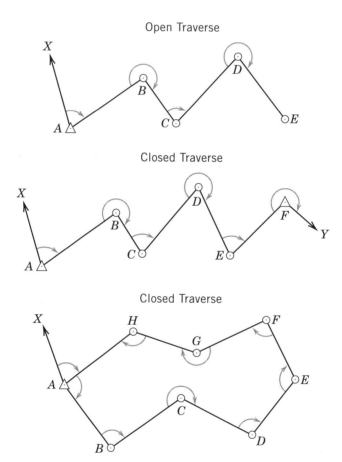

Open Traverse

Closed Traverse

Closed Traverse

28. Traverses

The Geodimeter and the Tellurometer were capable of producing comparable results; however, there were differences between them, the most obvious being the use of light waves versus radio waves. Among their other differences—differences typical of other instruments in these two categories—are:

Light Waves **Geodimeter**	**Radio Waves** **Tellurometer**
Control unit at one end of the line, reflector at the other end.	Similar control units at each end of line, one called the master and the other the remote.
Range limited by visibility to 30 to 50 km. Operator must be able to see between the terminals.	Possible range in excess of 100 km, even in poor visibility; i.e., can measure through haze, drizzle, etc.
Light waves affected only a little by atmospheric conditions.	Microwaves more susceptible to humidity changes.
Trained operator at the control unit, unskilled operator at the reflector.	Trained operator at each end.
No integral communication between the control unit and the reflector position.	Communication possible between the master and remote instruments on the measuring signals.

In general, the accuracies achieved were of the order of 1 in 250 000 (4 ppm) of the length of line; for example, an error of only 80 mm over 20 km. It is usual, however, to quote accuracy levels in the form ($\pm x$ ppm $\pm y$ mm) since part is attributable to the electronic configuration and part to the total length.

By the second half of the 1980s, accuracies in the submillimeter range were possible with the *Mekometer* at ranges of several kilometers, for example, $\pm$ (0.2 mm + 0.2 mm/km) up to 8 km.

Variations and improvements on these instruments continue to be made. However, many of the tasks they could be used for are now done by satellites in the GPS.

Traversing

The whole of surveying is involved with determining the relationships between points on the earth's surface. For large areas where there is reasonable intervisibility over great distances, then triangulation and trilateration will be used to form a skeleton of known points. Where the terrain is, for example, heavy jungle or bushland or unduly flat, another technique may be preferable. This

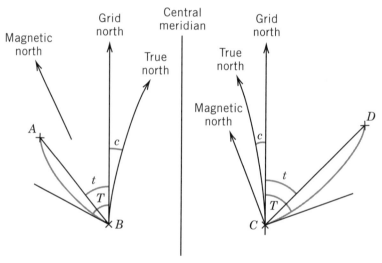

c = convergence
t-T = arc-to-chord correction

29. Different North Lines

method—called *traversing*—can be arranged in different ways according to the accuracy sought, the size, and the configuration of the area.

In essence (figure 28) traversing requires the measurement of the angles at a series of points together with determination of the distances between successive points. Angular measures invariably use the theodolite since these are available with resolutions down to 0.1″. The distances could be by invar tape in catenary (figure 26), steel tape on the ground, EDM equipment, or variations on these.

Networks

Network order　In each case, whether the surveyor uses triangulation, trilateration, or traverse the results are graded into first-order, second-order, and third-order (or primary, secondary, and tertiary) networks according to accuracy, area of coverage, and usage.

A *First-order* network is one for which the best available instrumentation has been used, the utmost care taken, and the most rigorous adjustment procedure applied. It is thus the most time-consuming and expensive framework but also the most accurate. Invariably, national surveys use first-order networks. The average length of line could well be 50 km.

Second-order work is the initial filling-in of the first-order network with closer points and to a less exacting standard.

A *third-order* network is yet a further breakdown of the secondary network, with coordinated points every 5 to 10 km.

Every national survey organization has its own criteria for each order based on local needs and circumstances.

Grids　For ease of location of points on a map the sheet is normally covered with a network of lines in two sets at right angles to one another. These are called *grid lines* and usually one set of lines is oriented more or less north-south. However, this is not essential, and in essence the lines can be rotated into whatever orientation the user requires.

In terms of the geodesy of this volume, it is usually the national grid system of a country that is of interest. Often it is stipulated that a survey scheme's coordinates must relate to the national system, which will almost certainly conform to a north-south orientation. At lower levels of surveying, quite arbitrary local grid systems can have any orientation the user thinks is appropriate. As noted elsewhere, coordinates on any one grid system can be converted to their equivalents on any other system, as long as there is a minimum number of points common to both systems.

Convergence　The relation between the "north" line on a grid and the direction to true north (the geographic North Pole) at any point is a variable. The difference between the two can be readily calculated and can amount to several degrees of arc (positive or negative) depending on the circumstances. This difference is known as *convergence* (figure 29).

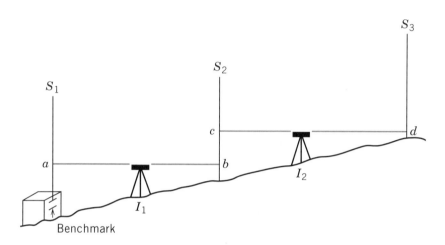

(a) Spirit Leveling

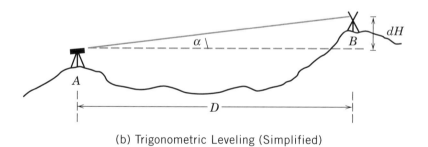

(b) Trigonometric Leveling (Simplified)

30. Leveling

Magnetic north   Yet another "north" to become familiar with is magnetic north. It is a slowly changing quantity and has to be calculated both as a function of the position of the point of interest and of the time (the nearest month or year) when the observation was made (see figure 29).

(t − T)   Just to complicate matters further, another correcting term is required. This is known as the (t − T) or *arc-to-chord correction*. A straight line between two points on the earth's surface will plot as a curve on a projection. This curve, or *geodesic*, diverges from the straight line by amounts that vary as a function of the position of the point of interest in relation to the central meridian of the projection. This divergence can be a handful of seconds, positive or negative, and so on occasion it has to be taken into consideration (see figure 29).

Spherical excess   Triangles formed on the earth's surface by well-separated points are not the same as those drawn for typical mathematical problems. In the latter, the angular sum always totals 180°, but that is not so with large triangles, which are essentially spherical triangles. The theoretical sum of the three angles in such a triangle is $180° + e$, where e can be a few seconds of arc and varies as the area enclosed by the triangle. As a rule of thumb, 76 square miles (195 sq km) equates to 1 second for e. Very roughly, this is an equilateral triangle with 10-mile (16-km) sides.

VERTICAL POSITIONING

Precise or Geodetic Spirit Leveling

Determination of the position of a point in the vertical plane above a defined datum, such as mean sea level, can take several forms. The most accurate is by *spirit leveling*. The instrument consists of a telescope set so that it can be very accurately leveled up in relation to a plumb bob (and hence to the geoid) and that can be rotated through 360° horizontally. It is used in conjunction with a graduated staff to find the change in elevation between a succession of points, as shown in figure 30a.

The height of the starting point, or benchmark, must be known. At the end of the sequence of observations a reading should be taken either to the same benchmark or to a different one if it is more convenient. This allows a ready check to be made for observational errors.

When the instrument is set up at I_1 and leveled (figure 30a) a reading a is taken on the staff S_1 at the benchmark. This allows calculation of the height of the plane generated by the line of sight (*plane of collimation*) when the telescope is rotated. A forward reading b is then taken on a second staff position S_2. The instrument moves on to I_2, and the same procedure is used for readings c and d and successive positions.

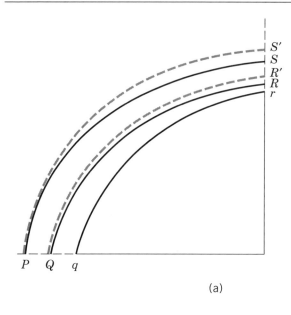

QR'; PS' = level surfaces - lines of constant orthometric height

QR; PS = equipotential surfaces - lines of constant dynamic height

(a)

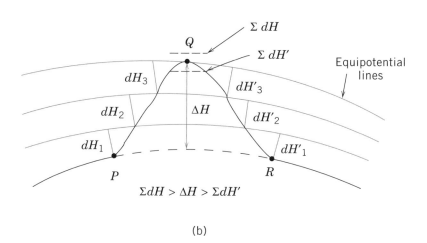

(b)

31. Orthometric and Dynamic heights

With this form of leveling, the distances S_1 to I_1, I_1 to S_2 and so on, are kept as nearly equal as possible and limited to some 30 to 50 m each.

This sequence is repeated as often as necessary to determine the heights of the required points. At all positions the rotating telescope is parallel to the geoid. Since the force of gravity increases as one goes from the equator to the poles, on leveling in such a direction the equipotential surfaces converge (figure 31a).

Orthometric, ellipsoidal, geoidal, and dynamic heights   For the survey practitioner, orthometric heights are the bread-and-butter form. Differences in orthometric height are the quantities obtained when using a normal level instrument. These differences are applied to the known value of a starting point to give *orthometric heights*. They are heights above the geoid. GPS gives height values related to an ellipsoid, which may be (but most likely is not) coincident with the geoid. Thus to turn GPS (ellipsoid) values into their equivalent orthometric values requires a correction called the *geoidal separation* (also known as *geoidal height* or *undulation of the geoid*).

If the ellipsoid and geoid happen to be coincident, then the separation is zero. On the other hand, if a different ellipsoid is used at the same point as before, there could be a noticeable separation (possibly many meters), either positive or negative. The sign indicates whether the geoid surface lies above or below the ellipsoidal surface at that point. Since the surfaces certainly are not parallel, the separation varies from place to place.

Often it is more important to know the difference between the separations at given points than the absolute values at those points.

There is a third form of height value—normally called *dynamic height*—that needs to be appreciated. Referring to figure 31, note that the linear distance Qq is greater than Rr, although R is said to have the same dynamic height as Q. A line of constant dynamic height is parallel to mean sea level along a line of latitude, but the lines coverage when progressing toward the poles. The orthometric height of Q above q is the length Qq. The dynamic height is measured as the work required to raise a unit mass from q to Q. Thus QR' is a line of constant orthometric elevation, while QR is a line of constant dynamic elevation.

Dynamic heights are thus defined in terms of potential, which is a function of the acceleration due to gravity and the distance of the point in question from the center of the earth. Equipotential lines (see figure 31) generally converge towards the poles. Dynamic height values are given in terms of *geopotential numbers,* where points of equal number lie on the same equipotential surface but will not have the same orthometric values.

The surface of a large lake is an equipotential surface. Leveling would show it to be (almost) level, but ellipsoid heights would generally show it to have a slope.

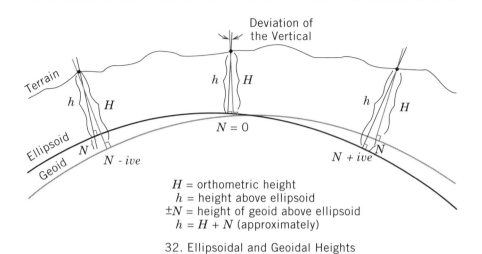

H = orthometric height
h = height above ellipsoid
$\pm N$ = height of geoid above ellipsoid
$h = H + N$ (approximately)

32. Ellipsoidal and Geoidal Heights

Orthometric heights are distances measured along the vertical perpendicular to mean sea level. The ellipsoid heights are measured along the normal to the ellipsoid. Thus heights found via GPS are ellipsoid values related to the normal to the ellipsoid. For comparison with spirit-level values, ellipsoid values have to relate to the normal to the sea level and the difference between the two surfaces (figure 32).

In the leveling process the bubble against which the instrument is leveled lies on an equipotential surface, but when the staff is read, it gives an orthometric value. An appropriate formula allows conversion from one to the other when the order of accuracy warrants it.

Note that while the traditional forms of heighting are related to the direction of gravity, the results from GPS are not so constrained.

Geoidal separation The geoidal separation can be determined with varying degrees of reliability. Where observational coverage has been good, contour maps of the values are often available, and an interpolation can be made for the required location. The accuracy may be of the order of 0.1 m depending on the part of the world involved. Other authorities provide computer models from which predicted values may be obtained.

Alternatively, an indication can be obtained by relating a selection of ellipsoidal results to known orthometric values in the same vicinity. A geodesist who does this, must take care in selecting suitable points.

For scientific purposes it is possible to observe astronomical positions along a particular line across a country and from the results determine, for a particular ellipsoid, the deviation of the vertical and geoidal separation. This is so time-consuming that it is seldom attempted, certainly not for standard survey operations.

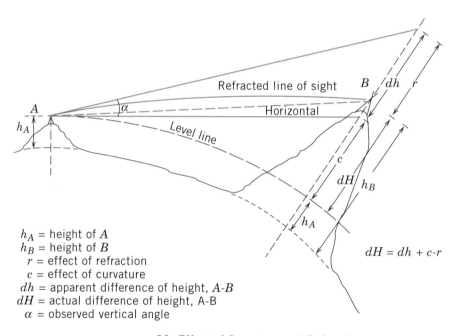

h_A = height of A
h_B = height of B
r = effect of refraction
c = effect of curvature
dh = apparent difference of height, A-B
dH = actual difference of height, A-B
α = observed vertical angle

$$dH = dh + c\text{-}r$$

33. Effect of Curvature and Refraction

Trigonometric Heighting

Trigonometric heighting is a method that can be adopted when it is necessary to determine the difference of height between points that are far separated or are difficult to level between, or where the required accuracy is not of the highest order.

Instead of a telescope mounted horizontally, *trigonometric heighting* involves using a theodolite to measure the vertical angles and combining them with horizontal distances (figure 30b) to find the difference in height. Since the earth is not a flat surface, the solution is complicated by the effect of the earth's *curvature,* which can be combined with the effect of atmospheric *refraction,* or bending of the light rays (figure 33). While the curvature is a definite correction as a function of the length of the line of sight, the refraction varies with the changing atmospheric conditions and so introduces an element of uncertainty into the result.

The combined effect of curvature and refraction is about 67 mm at 1 km, or 6.7 m at 10 km, or about 15″ of arc for 1 km in angular terms.

For this technique sight lengths can be as long as intervisibility allows. The height of Mount Everest was determined in this manner from numerous observations in the Indian Plains some 150 km away.

Variations on the technique are possible when both ends can be occupied simultaneously. It allows for better modeling of the refraction effect and improves the accuracy.

Barometric Heighting (Leveling)

Barometric heighting is a method used when lower-order accuracy is acceptable. It uses the principle of the change in pressure with elevation, first investigated and used by Bouguer while working on the Peru arc in 1738. In an ideal atmosphere there would be a direct relation between pressure and elevation, but in real life the situation is more complicated. Nevertheless, it is possible to use a pressure-measuring device for determining the difference of height. Using several such instruments in appropriate arrangements can improve the accuracy.

Originally a mercury barometer was required, which was difficult to transport and use satisfactorily. Nowadays aneroid barometers make the operation very straightforward. The aneroid consists essentially of a small box from which air has been exhausted. To its top surface is fixed a pointer and scale. The effect of changes in pressure on the box are magnified and indicated by the pointer. The units of measurement may be millibars of pressure, millimeters of mercury, or even direct graduation in terms of height units.

A variation on the mercury barometer no longer used is the hypsometer or boiling-point thermometer, which could be similarly related to height.

Echo Sounding

Besides methods to determine the heights of points on visible terrain, there are also depths below sea level to consider. In previous centuries the technique was to drop a weighted line over the side of the ship and record the amount of line paid out until the bottom was reached (see figure 44c). While this method was prone to various errors, it generally gave an acceptable indication of depth for the purposes of navigation. Nowadays far more information is required about the seabed, so sophisticated methods have to be used.

In particular, the *echo sounder* is a device for bouncing pulsed signals off the seabed and recording their return at the vessel. This can provide continuous coverage along a line as opposed to the discrete measures of a lead line. Variations on the basic principle allow recording of not only vertical depths but also a band to either side of the vessel's track. In this way complete coverage of selected areas is feasible. The higher the frequency of the signal, the shorter the wavelength and the narrower the beam width. This relation would be pertinent when accurate measurements of depth to specific points are of more interest than general indications of the seabed form and depth. Even so, accuracies much better than 1 in 200 are difficult to achieve.

An average value for the velocity of sound in seawater is 1500 m/sec, but this is a function of salinity, temperature, and density—all of which change with time and conditions as well as with depth.

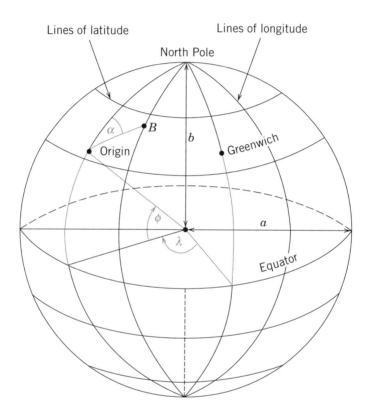

Lines of latitude   Lines of longitude

North Pole

ϕ = latitude
λ = longitude
α = initial azimuth

Geoid separation at origin = often zero.
a, b = semi-axes

34. Horizontal Datum

CHAPTER
5
Geodetic Systems

A *geodetic datum* is defined as any numerical or geometrical quantity or set of such quantities that serves as a reference or base for other quantities.

In geodesy two types of datum must be considered: a *horizontal datum,* which forms the basis for the computations of horizontal control surveys that consider the curvature of the earth, and a *vertical datum,* which elevations refer to. In other words, the coordinates for points in specific geodetic surveys are computed from certain initial quantities or datum parameters. Originally the horizontal and vertical datums were kept strictly separate, but today it is more usual to quote both at the same time.

This joint quotation results from the fact that orbiting satellites relate to the datum both horizontally and vertically. Previously there was no link or common factor between the two.

Because of the way in which surveying has evolved, there are a multiplicity of datums. In light of satellite geodesy, there would now be no need for more than one, but there is no likelihood of a full change since such a vast quantity of work corresponds to each separate datum. To convert all existing mapping information in just one country would take maybe a century. As with the property boundaries of centuries ago in the United States, so in years to come map users will still want to refer to the ancient systems for particular aspects. Thus a complete change would not eliminate a problem but rather add yet another variable to an already difficult inheritance.

HORIZONTAL GEODETIC DATUMS

A horizontal datum (figure 34) consists of the latitude and longitude of an initial point (called the *origin*), an azimuth for one line, the parameters for the ellipsoid chosen for the computations (one radius and the flattening), and the geoidal separation at the origin. Defining an ellipsoid does not on its own form a datum; it is only one of several required quantities.

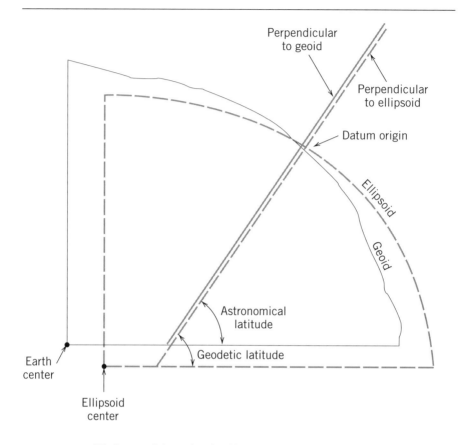

35. Datum Orientation for Single Astronomical Station

A change in any one of the parameters would affect all points computed on that datum. For this reason, while positions within a system are directly and accurately relatable, data such as distance and azimuth derived from computations between the positions of the same points on different datums will vary in proportion to the difference in the initial quantities.

Orientation of Ellipsoid to Geoid

Single astronomical position datum orientation (figure 35)   Selection of the reference ellipsoid provides the radius and flattening elements. The simplest means of obtaining the other three factors to establish the geodetic datum is to select a first-order triangulation station, preferably one located near the center of the triangulation network, to serve as the datum origin. Then the astronomical coordinates of the station and the astronomical azimuth of a line from that station to another control station are observed. An alternative to observing an azimuth would be to compute it by observing for the astronomical coordinates of the second station.

The observed astronomical coordinates and azimuth are adopted, without any correction, as the geodetic coordinates and azimuth of the datum origin on that particular ellipsoid. In addition, it is usual to assume that the geoid and ellipsoid are coincident at the datum. This means that the deflection of the vertical and the separation between the two surfaces are defined as zero at the origin. In this method of orientation, the normal to the ellipsoid is arbitrarily made to coincide with the plumb line at the datum origin.

Although the computed positions will be correct with respect to each other in this type of orientation, the entire net will shift with respect to the axis of the earth. This is not significant for local use of the positions but may introduce large systematic errors as the survey is expanded.

Note that although the deflection and undulation are defined as zero at the origin, deflections will occur at other positions within the network. Therefore, when we compare the geodetic latitude and longitude of any other point in the network with the corresponding astronomical latitude and longitude of that point, discrepancies will appear between the two sets of values.

A datum oriented by a single astronomical point may produce large systematic geoidal separations. The ellipsoid is not earth-centered, and its rotational axis is not coincident with the axis of the earth. The inconvenience of such an orientation is that the positions derived from different, astronomically oriented datums are not directly comparable to each other in any geodetic computation.

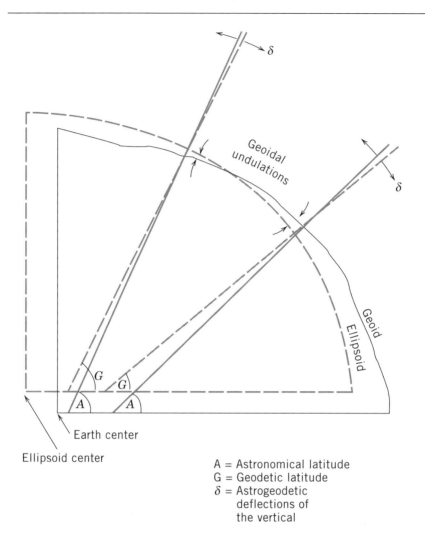

A = Astronomical latitude
G = Geodetic latitude
δ = Astrogeodetic
    deflections of
    the vertical

36. Orientation for Astrogeodetic Datum

Astrogeodetic orientation The deflections of the vertical at a number of Laplace stations (see below) can be used for a second type of datum orientation known as the *astrogeodetic orientation*. This orientation (figure 36) makes a correction at the origin that in effect reduces the sum of the squares of the astrogeodetic deflections at all the Laplace stations to a minimum. One of the Laplace stations in the adjustment is arbitrarily selected as the origin.

A *Laplace station*—named after the famous French mathematician Pierre-Simon Laplace (1749–1827)—is one at which astronomical observations are made for azimuth and longitude to control the accumulation of errors along the chain. If possible, they are located near the junctions of chains.

The longitude observation is made so that allowance can be introduced for the angle between the local vertical and the ellipsoid normal, which affects the azimuth observations. From this comes the noted Laplace equation that is a function of the difference between the astronomical and geodetic longitudes with the difference between the two azimuths.

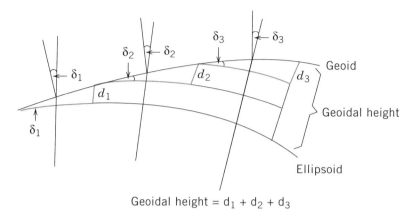

Geoidal height = $d_1 + d_2 + d_3$

37. Deflection of the Vertical

This deflection of the vertical (figure 37), or angle between the plumb line and the normal to the ellipsoid, is usually resolved into a north-south component (equal to the difference between astronomical and geodetic latitudes) and an east-west component (proportional to the difference between astronomical and geodetic longitudes.) The Laplace equation provides a means of reconciling the azimuth differences resulting from the use of two separate reference surfaces. Laplace equations are introduced into triangulation adjustments to control the azimuth and orient the ellipsoid. Therefore, instead of a zero deflection at the origin as with a single astronomical position, there is a particular deflection of the vertical. Similarly, the geoidal separation (undulation) can be determined at the origin and the ellipsoid reoriented provide a best fit for the ellipsoid and the geoid in the area of the Laplace stations used. Consequently astrogeodetic-oriented datums can be applied to larger areas than those oriented by a single astronomical position.

This, for example, was the arrangement applied a few years ago in Australia. Using an adopted origin with coincident astronomical and geodetic coordinates, the survey networks were preliminarily computed. This provided a set of deflections of the vertical from which a mean value was obtained. Application of this mean value to the origin allowed a recomputation that gave a better overall fit for the whole continent.

In general, a best-fitting ellipsoid can be chosen for a national geodetic network such that the undulations are no more than 10 m.

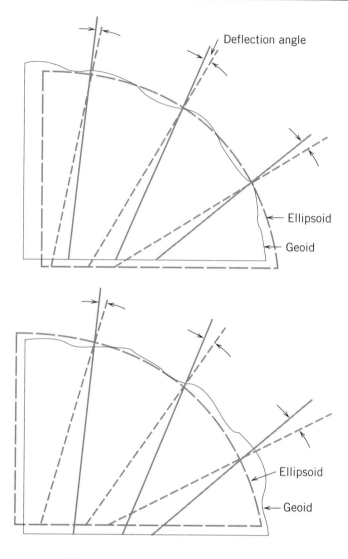

38. Astrogeodetic Deflections Are Relative

The astrogeodetic orientation has the disadvantage that the deflections of the vertical remain relative (figure 38). If the ellipsoid is changed the deflections of the vertical will also change. Second, it is necessary to assume a specific orientation of the reference ellipsoid with respect to the geoid before computing the astrogeodetic deflections. The orientation is fixed by the initial values of the datum origin from which the geodetic coordinates were computed. Any change in these quantities changes the deflection of the vertical at each point. Consequently, the astrogeodetic deflection of the vertical depends upon a specific geodetic datum, and the use of geodetic data developed by this method is limited to relatively small areas.

The angle between the *astronomical zenith* (which relates to the geoid) and the *geodetic zenith* (which relates to the ellipsoid) equals the *astrogeodetic deflection*. This angle also equals the angle of intersection between the geoid and ellipsoid surfaces, so it is possible to compute the separation in increments from the deflection values(see figure 36).

Satellite Datums

The datum used for satellite observations is defined in a different way. It is based on the positions given for several tracking stations, the geopotential model selected, and various constants that do not appear in traditional datums. These include the velocity of light, the rate of rotation of the earth, corrections for clock and oscillator rates, the gravitational constant, and the earth mass. Such a system has as its origin the earth's center of mass and uses the same coordinate system as that used for the precise ephemerides (see page 145). It is usual also to quote an ellipsoid of reference so that geographic coordinates can be obtained.

Discrepancies between Datums

In areas of overlapping geodetic triangulation networks, each computed on a different datum, the coordinates of the points given with respect to one datum will differ from those given with respect to the other. The differences occur because of the different ellipsoids used and the different deflections of the vertical and geoid separations at the datum origins. The latitude and longitude components of the absolute deflection at the initial points result in a parallel shift between the systems. Such a shift is due to the fact that the minor axes of the various reference ellipsoids do not coincide with the rotation axis of the earth. In addition, deflection errors in azimuth cause a relative rotation between the systems. Finally, a difference in the scale of the horizontal control may result in a stretch in the corresponding lines of the geodetic networks.

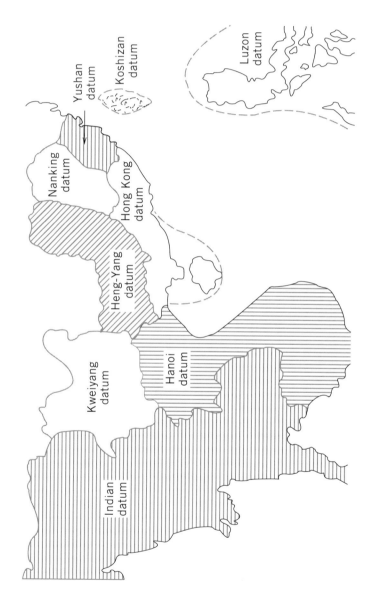

Yushan datum

Koshizan datum

Luzon datum

Nanking datum

Hong Kong datum

Heng-Yang datum

Kweiyang datum

Hanoi datum

Indian datum

39. Example of Many Datums in Southeast Asia Area

The discrepancies are generally larger between datums oriented by a single astronomical point than between those with an astrogeodetic orientation.

In view of the initial point translation, the azimuth rotation, and the horizontal scale differences, any attempt to correlate the geodetic information from one datum to another, unconnected datum is quite impossible. Regardless of the accuracy of the individual datums within themselves, there is no accurate way to perform distance and azimuth computations between unconnected geodetic systems.

With the development of both intermediate and long-range weapon systems, geodetic problems have become more critical than ever before. To satisfy military requirements, it is necessary to provide detailed cartographic coverage of areas of strategic importance and to accomplish geodetic computations between these areas and launch sites, which are often on unrelated datums. Both of these requirements necessitate unification of major geodetic datums by one or a combination of existing methods.

Datum Connection

There are three general methods by which horizontal datums can be connected. The first method is restricted to surveys of a limited scope and consists of systematic elimination of discrepancies between adjoining or overlapping triangulation networks. This is done by moving the origin and rotating and stretching the networks to fit each other. The method is generally used to connect local surveys for mapping purposes. Known as datum transformation it can only be used where control exists for common points in the different systems. A minimum of two such points is necessary, but it is preferable to have a spread of common points over the area of interest. It is then possible to do a least-squares best fit between the two sets of values.

To relate larger areas to one another, common points can be used if they exist; in addition, extra common positions can be introduced through satellite observations. Using some of the methods described in Chapter 8, the geodesist can interrelate satellite observations from critical points in each geodetic network and so calculate the necessary transformation parameters. In earlier years this same form of connection could be achieved through solar eclipses, star occultations, and use of the moon-position camera.

The third form of datum connection is a gravimetric method, described on pages 121 and 127.

Coordinate transformations can be in either two or three dimensions depending on the requirements and the common data available.

Datums before World War II

By 1940 every technically advanced nation had developed its own geodetic system to an extent governed by its military and economic requirements. Some systems developed through the expansion and unification of existing

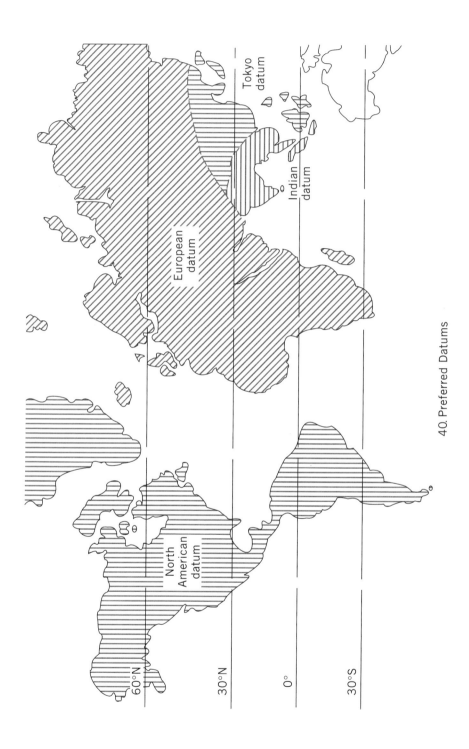

40. Preferred Datums

local surveys, and others by new nationwide surveys replacing outdated local ones. The former of these two approaches is not to be recommended since "working from the part to the whole" can propagate and enlarge inherent errors and is contrary to the standard survey maxim of always "working from the whole to the part." Normally, neighboring countries did not use the same geodetic datum. There was no economic requirement for common geodetic information, and the use of common datums was contrary to the military interests of each country. The only surveys of an international nature based on the same datum were the few measurements of long arcs accomplished for the purpose of determining the size and shape of the earth. An example of this was the Struve arc from the River Danube to the North Cape. This meridian arc was measured by triangulation during the period 1815 to 1855; it stretches 25°20' in latitude or over 2800 kms, and it passes through ten countries roughly following the 26° E line of longitude.

International boundary surveys required coordinates of marks to be in the systems pertinent to the countries involved, but this could be achieved without full conversion into one united system. The net result was that there were many different surveys of varying size that differed from each other to a remarkable extent. The national maps based on the surveys also differed widely. Figure 39 illustrates the confusion of geodetic information in Southeast Asia.

As military distance requirements increased, positioning information of local and even national scope became unsatisfactory. The capabilities of the various weapon systems increased until datums of at least continental limits were required.

Preferred Datums

By 1940 the best solution was to establish for both military and scientific purposes a set of preferred datums for selected areas and adjusting all local systems within each area to it. The North American, European, Tokyo, and Indian datums were selected for this purpose (figure 40). It did not mean that each country had to convert all of its coordinated material to the preferred datum for everyday use. It was only information of a primary character and of scientific and military significance that obtained a second set of values.

North American Datum (NAD) Before 1927 a *U.S. standard datum* was based upon a meager triangulation network. As new networks were added, the magnitude of the inconsistencies became objectionable.

As a result, a general readjustment was made of the whole system, and the outcome was known as the *NAD 27 datum*. The origin is at Meades Ranch, Kansas, lat. 39°13'26.686", long. 261°27'29.494". The datum was computed on the Clarke 1866 ellipsoid, which was oriented by a modified astrogeodetic method. The system incorporates Canada, Mexico, and the West Indies with a Central and South American connection. Consolidation of the various Central and South American surveys has been carried out by the Inter-American Geodetic Survey.

As before, inconsistencies became apparent as new data was added and higher reliabilities were required between relative coordinate positions. Where an accuracy of 1 in 100 000 was required, some parts had only 1 in 15 000. Hence errors and distortions were introduced.

Many years of effort based upon a vast increase in data, improved computational techniques, and the advent of satellite positioning systems went into the redefinition. The result was the adoption of the 1980 international ellipsoid values (see page 31). The new datum became known as *NAD 83* and had a geocentric origin.

European Datum (ED)   The initial point of *European datum* is located in Potsdam, Germany, at the Helmert Tower, lat. 52°22'51.45" long. 13°03'58.74". Numerous national systems were joined into this large datum based upon the international ellipsoid, which was oriented by the astrogeodetic method. The Army Map Service connected the European and African triangulation chains and filled the remaining gap of the African arc measurements from Cape Town to Cairo. Thus all of Europe, South Africa, and North Africa were molded into one great system. Through common survey stations it was also possible to convert data from the Russian Pulkovo 1932 system to the European datum; as a result the European datum includes triangulation as far east as the 84th meridian. Additional ties across the Middle East permitted connection of the Indian and European datums.

ED 50 covered various countries on mainland Europe, but in 1963 a connection was made across the English Channel to join the United Kingdom to the same datum. Initially ED 50 was used mainly for scientific purposes, but later it became useful in such operations as the oil industry in the North Sea. This stretch of water lies between Scandinavia and the British Isles, and so a common system was of particular advantage. In the 1980s and 1990s the Channel Tunnel connection has been an opportunity for cooperation between the national survey organizations of France and Great Britian and for use of the common datum.

Tokyo datum   The third preferred datum, the *Tokyo datum,* has its origin in Tokyo at the Observatory, lat. 35°39'17.51", long. 139°44'40.50". It is defined in terms of the Bessel ellipsoid and oriented by means of a single astronomical station. By triangulation ties through Korea, the Tokyo datum is connected with the Manchurian datum. Unfortunately, Tokyo is situated on a steep geoid slope, and the single station orientation has resulted in large systematic geoid separations as the system extends from its initial point.

Indian datum   The *Indian datum* is now accepted as the preferred datum for India and several adjacent countries in Southeast Asia. It is computed on the Everest ellipsoid with its origin at Kalianpur in central India, lat. 24°07'11.26", long. 77°39'12.57".

Derived in 1830, the Everest ellipsoid is the oldest of the ellipsoids in common use and is noticeably out of sympathy with more recent determinations. As a result, the datum cannot be extended too far from the origin since very large geoid separations will occur. For this reason and the fact that the ties between local triangulations in Southeast Asia are typically weak, the Indian datum is probably the least satisfactory of the preferred datums.

Thus it can be seen that 4 quite different ellipsoids are used. Referring to page 31, their parameters are:

	a	$1/f$
Clarke 1866	6 378 206	294.978
International (Hayford)	6 378 388	297.000
Bessel	6 377 397	299.153
Everest	6 377 276	300.802

As figure 41 shows, wide variations of this magnitude can not be readily reconciled into one system without large distortions and deviations.

Other Arrangements

For military distance and direction problems, limited to continents or smaller areas, the preferred datums were initially satisfactory. However, while they were improvements over the limited national datums, they had serious deficiencies that prevented them from providing the geodetic information required for intercontinental ballistic missiles.

At one stage a North Atlantic Hiran tie permitted connection of the European datum and the North American datum. This however, did not completely solve the problem since both these datums are relative. While in each case the chosen ellipsoid gave an adequate fit in the area of the origin, as already shown, neither provided a good fit for the entire earth. In addition, the process of connecting various datums by means of intervening datums or triangulation/trilateration ties allowed values to accumulate that do not always agree with newly observed data. The surveys joining India to the European and Tokyo datums presented just such a major problem; the requirement for long-range geodetic information could not be considered solved by the preferred datums.

The Hiran system (mentioned above) was one of the early improvements on the *Shoran system*, which was the first application of radar techniques to surveying. The principle used was similar to that of modern EDM equipment. However, by mounting the equipment in an aircraft much longer lines could be measured, and some 500 km proved feasible. The Hiran system permitted the use of even longer lines and achieved an accuracy of 1 in 100 000 (see figure 53).

Choice of ellipsoid   While it might seem obvious that it would be preferable to have just one ellipsoid for worldwide use, this is not necessarily practical. As figure 41 shows, an ellipsoid that fits well in one part of the world may diverge by a noticeable amount elsewhere. The effect of this was illustrated in Australia, where, had they adopted WGS 72, the divergence between the ellipsoidal and geoidal surfaces would have given unacceptable scale errors. Corrections varying

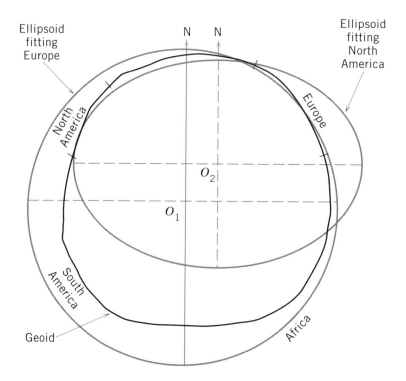

41. Fitting Different Ellipsoids to Different Areas

from +6 to −12 ppm would have had to be applied by everyone in the distance measuring business many of whom would not have understood what they were doing, and hence would quite likely ignore the corrections.

An African Datum   Recently new work has been done to assess the options for choosing a suitable datum for the whole of Africa. Earlier attempts, completed some 10 years ago, have not yet been implemented because the many changing boundaries in the continent make a unified approach important. Doppler data relating to 185 stations throughout the continent, together with other available information, have been used to find the best-fit parameters under a variety of situations. The results vary considerably, and much further work is necessary before any one arrangement can be accepted.

VERTICAL DATUMS

Just as horizontal surveys refer to specific original conditions (datums), so vertical surveys also relate to an initial quantity or datum. As already discussed, elevations are based on the geoid because the instruments used either for spirit leveling or trigonometric leveling are adjusted so that the vertical axis is coincident with the local vertical.

As with horizontal datums, there are many small discrepancies between neighboring vertical datums. Although elevations in some areas relate to surfaces other than the geoid, there is never more than 2 m of variance between leveling networks based on different mean sea-level datums.

In European areas there are fewer vertical datum problems than in Asia and Africa. Extensive leveling work has been done in Europe, and practically all of it has been referred to the same mean sea-level surface. However, in Asia and Africa the situation is entirely different. In some places there is precise leveling information based on mean sea level; in other areas the zero elevation is assumed elevation that sometimes has no connection to any sea-level surface. China is an extreme example of this situation, in that nearly all of the provinces have their own independent zero reference. There are no reliable vertical data for as much as 75% of the area of Asia and Africa.

As long ago as the Peruvian arc measure of 1738 there have been difficulties in connecting to sea level. Only after several years of trying was Bouguer able to tenuously connect the triangulation scheme of the high Andes to sea level and make the appropriate corrections to the observations and calculations.

The vertical reference system of the United States was fixed in 1929 from some 100 000 km of leveling on the North American continent. Twenty-six tide-gauge values were held fixed in the overall adjustment. A new vertical control network is expected to be completed in the near future.

The datum in the UK was determined in 1921 at Newlyn while that in Australia was determined in 1971 on the basis of a national levelling net adjustment that incorporated a large number of tidal stations.

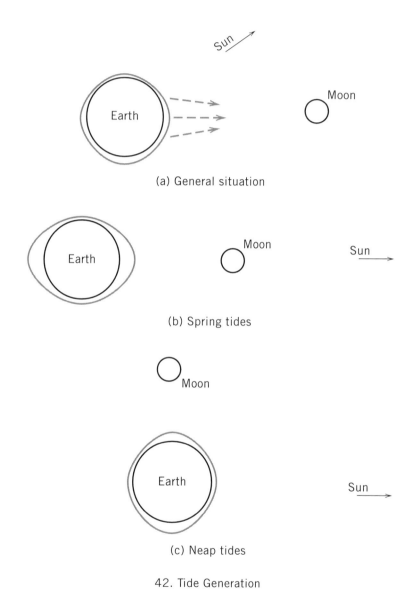

(a) General situation

(b) Spring tides

(c) Neap tides

42. Tide Generation

Tides

Whereas for horizontal datums there is an infinite choice of datum positions when it comes to the datum plane for heights, the choice is simpler. Throughout the world the oceans present an obvious choice as a datum, and that is what is often chosen, but the situation is not quite as simple as it would seem. Since the fourth century B.C. it has been appreciated that the sea level rises and falls and that the amount of rise and fall varies throughout the year. (Tidal variations in the Mediterranean area were unimpressive and it was not until wider exploration that their regularity became accepted). What value then should we take as the sea-level position when it is to be a datum?

What causes tides   Consider first the forces that produce the tides. It was not until the time of Newton that the forces producing tidal variations were properly understood. He showed how the tides were in direct agreement with his law of gravitation. While the sun revolves in relation to the earth in about 365.25 days, the moon does so in about 27.3 days. Between them these bodies exert forces upon the earth in keeping with Newton's law that every particle of matter attracts every other particle with a force proportional to the mass of each and inversely proportional to the square of the distance between them. Since the sun and moon neither orbit in circles nor lie in similar orbits, any effect they have varies from day to day according to the relative position of both bodies to the earth at any particular time.

The orbits of the sun and moon are ellipses, so there are positions of nearest and farthest approach for each body. If we consider each body separately and assume that the earth is completely covered with water, then the water height will rise at places on the earth approximately where the line normal to the body cuts the water surface. This will apply on both sides of the earth (figure 42). Similarly, there will be depressions of the surface at positions 90° from the high points. As the earth rotates, these positions will move around it. At the same time, as the body in its orbit varies in its distance from the earth, its attractive effect changes and the height to which the water surface rises varies accordingly.

Since there are two positions of both high and low water level, at any point on the earth's surface there will usually be two high water levels and two low water levels every day. As stated above, these positions lie approximately on the intersection of the normal to the earth's surface and the body. They are only approximations because of the time lag caused by the effect of friction between the water and the ground surface over which it moves. The positions of the sun and moon in relation to the equatorial plane also cause variations in the times of occurrence of high and low water levels and their magnitudes.

In general, the moon's effect is more than twice that of the sun. There are times during the year when the two bodies tend to act together and other times when they act against one another. Thus we can see that many parameters are involved in finding the position of the water surface at any time. In fact, there are even longer-period variations, such as 18.6 years due to movements of the position of the moon relative to where its path crosses the ecliptic.

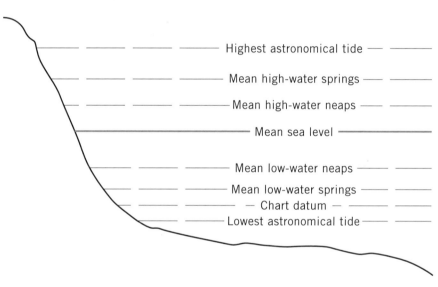

Highest astronomical tide

Mean high-water springs

Mean high-water neaps

Mean sea level

Mean low-water neaps

Mean low-water springs

Chart datum

Lowest astronomical tide

43. Selection of Ocean Levels

Since quarter-diurnal, semidiurnal, diurnal, monthly, annual, and 18-year variations need to be combined with the effect of the topography of the coasts and unpredictable fluctuations, it is no wonder that choosing a datum is difficult.

Mean sea level To obtain the best datum position at any coastal location, continuous observations of the sea level should be taken over a period of at least 18.6 years and the results averaged. This would give the mean sea level at that location. Such a length of time could be impractical, and approximations of varying accuracies are obtained by averaging regular observations over shorter periods such as a day, (actually the average length of a tidal day is nearly 25 hours), a month, or a year.

The accepted constituents of the variations can range in number according to the detail and accuracy required in the overall result. Up to 63 such components are sometimes referred to, but that would only be for exceptional investigations. Several groupings within these interact with each other to provide repetitions over cycles ranging from a few hours to many years.

Monitoring Recent environmental worries over gas emissions have suggested that sea level may be rising steadily. Obviously, for many reasons such a change requires careful monitoring. The trend may be only a millimeter or two per year, but aggregated over some years this can become a disastrous amount—so much so that in various parts of the world continuous monitoring of the sea level is an ongoing concern. For example, many of the islands in the Pacific Ocean are low-lying, and any noticeable sea level rise would cause not only inundation but perhaps annihilation of some islands. Because of this, a network of 11 monitoring stations has been set up around the islands stretching from Papua New Guinea at about long. 148° E to the Cook Islands at 162° W, 20° S and to the Marshall Islands at 10° N or over an area 50° E-W and 30° N-S.

The system to be set at each of these stations will be the latest state-of-the-art multimeasurement arrangements to record all manner of sea parameters and to transmit the results to the master processing station via a geostationary Japanese satellite. The system will be capable of taking readings every 2 seconds with the results averaged in 6-minute blocks.

While all these measurements may be good to 1 mm, remember that the earth's surface itself is moving. Hence all such stations have to be located on a single reference system—the (International Earth Rotation Service, Terrestrial Reference Frame (IERS TRF)—and monitored over many years before definite sea-level movement can be confirmed. For this, one of the best methods of relative positioning over thousands of miles is very long baseline interferometry (VLBI).

Definitions Mean sea level can be defined in two different ways. It is the equipotential surface that the oceans would have if left undisturbed. Alternatively, at any individual point on the seacoast it can be defined as the average of the constantly changing water level over a period of time. This could be better termed *local mean sea level.*

Many alternative positions of the water surface (figure 43) have their uses for navigation and other purposes. In particular, there are the values of high

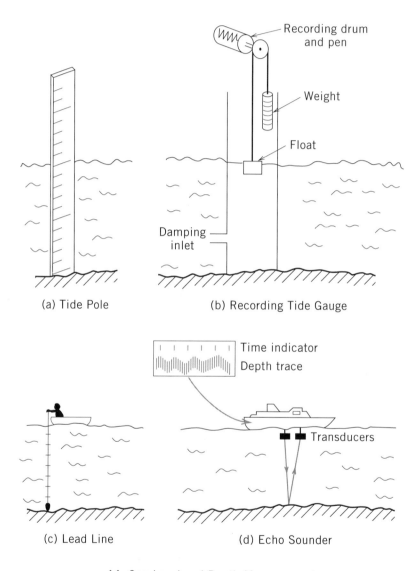

(a) Tide Pole

(b) Recording Tide Gauge

(c) Lead Line

(d) Echo Sounder

44. Sea-Level and Depth Measurement

and low water at spring and Neap tides. These are the times of greatest and least range in the diurnal tide heights and occur about every 2 weeks. Their cause is the relative alignment of the sun, moon, and earth. When all three are in line—at the full or new moon—the tides are highest (springs tides), and when the sun and moon are widest separated—at the first and third quarters of the moon—their effects tend to cancel and give the lowest (neap) tides.

Of particular importance is the value taken as datum for indicating depths on charts. Theoretically, it is a position below which the lowest water level never falls. This approximates what is called *lowest astronomical tide,* or the position below which predictions suggest that the water will not fall under normal conditions.

Different parts of the world have different definitions of their vertical datum, but all approximate to that outlined above. Any one country should have one main datum, with any subsidiary ones determined in the same manner as the main one and connected to it by geodetic leveling. Otherwise if leveling operations started from one datum and closed on to another, there would be undue discrepancies if the basic criteria for their establishment had been different.

Recording How can the mean sea level be recorded? The simplest method is to plant a graduated pole (figure 44a) in the water at the required location. Then at regular intervals—say every hour—take readings at the water level on the pole. This process has the obvious disadvantages of being tedious, not particularly accurate, and occurring at discrete intervals of time rather than being continuous.

At the other extreme is the automatically recording tide gauge (figure 44b). This has continuous recording by pen and graph paper affixed to a revolving drum. One revolution of the drum usually equates to 24 hours. The pen is linked to a float so that as the float moves up and down, so does the pen. Various forms of damping device are incorporated to smooth out the surges of the waves and latest developments incorporate automatic recording onto a data storage and computer facility.

One notable modern advance in this area is the air-acoustic sensor system connected to a microprocessor data acquisition system. Resolutions to the millimeter are readily obtained. The principle behind the system is that a transducer, set in a tube, emits a pulse from a fixed position toward the water surface. At a known position along its path, some of the pulse hits a bar placed in the path and is reflected back. The rest of the pulse continues to the water and is reflected from its surface. The first part of the pulse—that reflected from the bar—travels over a known path length, and is used to calibrate the system for the velocity of sound. Hence the value found for the overall reflection can be used to determine the position of the water surface below the transducer.

With either of these forms of tide gauge or with the variations on these themes (figure 44a,c), it is particularly important to select the site carefully.

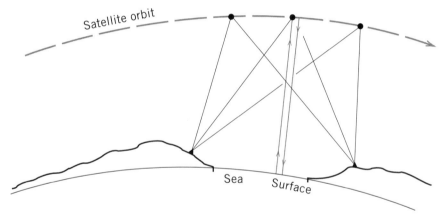

45. Satellite Altimetry

It should be near the open sea but not unduly affected by abnormal wind and storm conditions. If the gauge is to be monitored for many years, then it should be at a site that does not silt up quickly. Any change of the seabed in this way can affect the readings.

In all cases the tide-gauge zero mark should be accurately located in terms of how far it is below a permanent first-order benchmark to which subsequent level measurements can be easily related.

Coriolis Effect

The rotation of the earth causes what is known as the *Coriolis effect*. Besides having an influence on tidal movements, it also affects the movement of artificial satellites. Since water is stationary in relation to the earth, as the earth rotates, so the water near the equator is rotating at some 1400 km/hr while toward the poles, the rotation slows. If various forces then act to move a particle of water toward the east in relation to the earth, the particle will actually begin to drift toward the south because it would be moving faster than normal for its latitude. If we followed the particle for awhile, we would observe this tendency moving it steadily in a clockwise direction if north of the equator and counterclockwise if to the south of the equator.

In a similar manner a satellite launched toward due north from near the equator would have both its own launch velocity and the velocity due to the rotation of the earth. This combination of components would cause it to drift east of its initial due-north direction and to trace out a path over the earth's surface that curved to the east. As with the water, so the satellite to the south of the equator would appear to drift in the opposite direction.

Satellite Altimetry

A few satellites are now able to measure heights above the mean wave surface of the oceans. For this there is no possibility of a "ground" transmitter or receiver at the required height position, so the measure has to be made in terms of the time a signal takes to make the round trip from the satellite to the sea and back again. From a sequence of such signals a mean value is obtained; however, as with echo sounding to the ocean floor, it is no good having a depth if its position is unknown.

In the case of satellite altimetry (figure 45), it is the satellite's position that has to be accurately determined for the time of measurement. This is achieved by observing the satellite by either laser ranging or Doppler techniques that allow determination of the orbit. Then from knowledge of the orientation of the transmitter/receiver signals, the depth to the wave surface can be calculated.

The value found will not be mean sea level because of tidal, atmospheric, and other effects; nevertheless, it provides useful data for the determination of mean sea level and, more particularly, for the geoid surface, which, strictly speaking, differs from mean sea level by small amounts. This arises because the acceptance of mean sea level as an equipotential surface is affected by a variety of perturbing effects that cause a meter or so in variation.

Transformations

With such a multiplicity of systems worldwide it is essential to be able to convert (or transform) values on one system to their equivalent on another system. If one were thinking only of a small area (few square miles) over which the values are spread, then a very simple *two-point transformation* will suffice. This requires only two points common to the two coordinate systems in question. From such a minimum of data it is possible to allow for differences between the systems that involve scale, unit of measure (for example, from a system in imperial units to one in metric units), orientation, and origin.

In such an arrangement, however, there are no redundant observations, no checks on gross errors. Two simple formulae are all that is required to convert the values of any number of points within that limited area.

However, in the more appropriate geodetic setting, seven parameters are required to cover a scale factor, rotation around the three axes, and movement of the origin in three directions.

It is possible to convert between any two coordinate systems whether their positions are defined in cartesian form (that is, east, north, and height) or in geographic form (latitude, longitude, and height). References such as that by Soler and Hothem (1988) give a good lead into this subject.

Typical datum shifts to WGS 84 are:

From	dX	dY	dZ	Scale
OS(SN)80	+364.705	−109.420	+429.003	−1.2 ppm
WGS 72	0	0	+ 4.5	+0.23
NAD 83	0.42	0.95	− 0.62	−0.14
NAD 27	− 4	166	183	0.37
Indian	227	803	274	6.59

It can be seen from just these few examples that shifts can range from nothing to many hundreds of meters.

There are various methods by which the transformations can be executed. All are beyond the scope of this volume, but Deakin et al. (1994) for example, compare the results from least-squares collocation with the three-parameter method of translations only and the seven-parameter one of translations, rotation, and scale factor. The differences between the results in their particular example and the published values for 11 stations were, in all except three cases out of the 66, less than 1 m. No particular conclusion can be drawn as to which is the preferable approach, and no doubt this is an area where discussion will continue for some time to come.

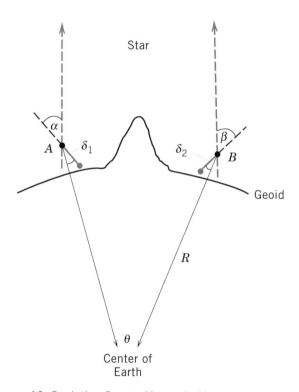

46. Deviation Due to Mountain Mass

CHAPTER
6
■

Physical Geodesy

This chapter introduces a branch of geodetic science that is closely related to geology and geophysics. The basic distinction between *physical geodesy* and *geometrical geodesy* is that the former utilizes measurements and characteristics of the earth's gravity field, in addition to theories regarding this field, to deduce the shape of the earth and, in combination with arc measurements, the earth's size; the latter, as we have seen, is concerned with the methods of measurement and computation on a curved surface and the determination of the parameters of the reference ellipsoid. With sufficient information regarding the earth's gravity field, it is possible to determine geoidal undulations, gravimetric deflections, and the earth's flattening.

Pierre Bouguer in 1738 endeavored to quantify the deflection of the vertical during the Peru survey expedition. He made observations near Mount Chimborazo in the Andes but conceded that the inaccuracies in his method were probably equivalent in magnitude to the quantity he was trying to determine. Thus his results were of little use other than to pave the way for future investigators.

In 1774 Nevil Maskelyne, Astronomer Royal at Greenwich, attempted to measure the effect of attraction in relation to Mount Schehallion in Scotland, an east-west ridge standing 600 m above a fairly level area. Figure 46 shows attractions δ_1 and δ_2 at points A and B together with zenith distances α and β to a star. From knowledge of R (radius of the earth) and θ (the angle subtended by A and B at the center of the Earth) found by survey methods, the amount of deflection could be calculated. Maskelyne found a residual displacement of 11.7″ of arc as the attraction and concluded that the earth had a mean density 4.867 to 4.559 times that of water. (The present-day accepted value is 5.52.)

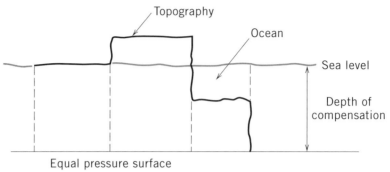

Equal pressure surface

(a) Pratt's Hypothesis

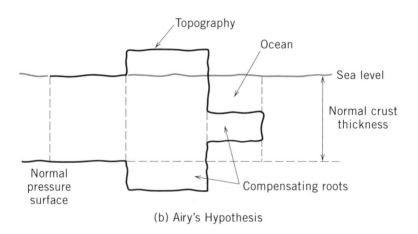

(b) Airy's Hypothesis

47. Isostatic Hypotheses

ISOSTASY

Following the efforts of Bouguer there were a number of attempts to develop a theory for the distribution of masses within the earth that would satisfy observations. Was there in fact a deficiency of material below a mountain? Were there large, dense, hidden masses scattered through the earth?

The theories of the early 19th century allowed estimates to be made of the attraction caused by the Himalaya mountains. One notable personality was Archdeacon John Pratt, who calculated theoretical values for the possible attraction. However, when the Survey of India was observing its great triangulation, the differences between astronomically observed and geodetically computed locations gave the attraction to be only about one-third the theoretical value.

To explain this discrepancy, both Pratt and the then Astronomer Royal at Greenwich, Sir George Airy, put forward conflicting ideas (figure 47) based on their thoughts regarding the distribution of masses within the earth. Pratt (in 1854) suggested that the crust of the earth was uniformly thick below sea level but that the density varied to account for the topography. Airy (in 1855), on the other hand, considered that beneath mountains the earth crust dipped downwards like the root of a tooth, so that down to some particular level in the earth's crust the total mass per unit area would be the same.

These theories of the earth's crustal structure now form part of the science of isostasy, although it was some years after the time of Pratt and Airy when that term was introduced. Its basic concept is that the outer part of the earth adjusts the distribution of its component masses to reach equilibrium. Thus, to obtain reasonably accurate theoretical values of the gravitational field throughout the earth, we must consider the effect of isostasy.

In using the earth's gravitational field to determine the shape of the earth, the gravimetrist (someone who measures gravity) measures the acceleration of gravity at or near the surface of the earth. It is interesting to compare the acceleration measured by the gravimetrist and the acceleration experienced in an airplane. In the latter, an acceleration is simply called a g force and is measured by a g meter. A g factor of unity is used to indicate the acceleration due to the attraction of the earth and is considered a neutral condition. The gravity unit used and measured in geodesy is much smaller: A g factor of 1 is approximately equal to 1000 gal. Thus in geodesy, where the milligal (see p. 115) is the basic unit, we are dealing with variations in acceleration equal to 1 millionth of a 1-g aircraft acceleration. In fact, modern intruments allow the ready resolution of 0.01 mgal.

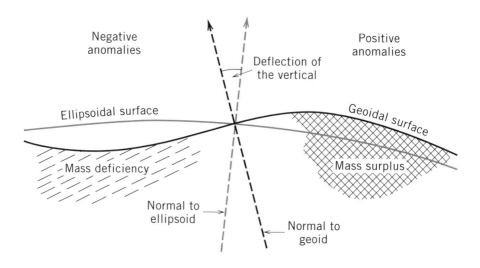

48. Effect of Mass Anomalies

GRAVITY ANOMALIES

By assuming the earth to be a regular surface without mountains or oceans, with no variations in rock densities or in the thickness of the crust, a theoretical value of gravity can be computed for any point by a simple mathematical formula. This theoretical value represents the combined effect of the force of the earth's attraction due to gravitation and the centrifugal force due to the rotation of the earth. The theoretical value of gravity at a point on the ellipsoid's surface depends both on the size and shape of the ellipsoid and on a value, computed from observational data, that is accepted as the theoretical value of gravity at the equator. It varies only with the latitude of the observation point if the figure of the earth is taken to be an ellipsoid of revolution, although some experts have tried to introduce a longitude term as well. While there are several formulae for computing the theoretical gravity, the one most commonly used today, resulting from modern evaluations of the earth's parameters, is based on the Geodetic Reference System 1967 (GRS 67). Before then the formula used was based on the International Ellipsoid (1924). The *gravity formula 1967* is given by

$$\gamma = 978\ 031.85\ (1 + 0.005\ 278\ 895\ \sin^2 \phi - 0.000\ 023\ 462\ \sin^4 \phi)$$

where

ϕ = geodetic latitude
γ = normal gravity in milligals (mgal) at the ellipsoidal surface

The differences between theoretical and observed values are termed anomalies. Used in this sense, an *anomaly* is a deviation from the normal and can be used either for a single point or to describe a regional effect (figure 48). To make use of the anomalies, the observed gravity must be reduced to a common frame of reference—the geoid or mean sea level. In an ideal situation all anomalies would be zero—that is, theoretical values would equal observed values—but this does not happen. The calculation of possible causes for these

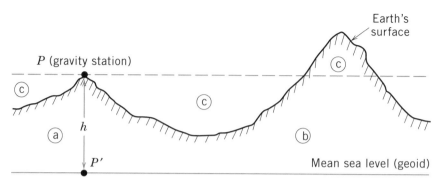

(a) Height h of observation point P
(b) Mass between mean sea level and height h
(c) Effects of terrain surplus and deficiency

49. Factors Involved in Reducing Observed Gravity

anomalies can be achieved according to different assumptions. The reductions may take into account the elevation above or below sea level, account for the terrain surrounding the point, and the assumed structure of the earth's crust (figure 49). The position of the point must be established with sufficient accuracy to meet the gravity survey requirements. There must also be accurate enough topographic information of the distant as well as surrounding area to satisfy reduction needs.

Three particular methods of assessing for the anomalies are worthy of mention here.

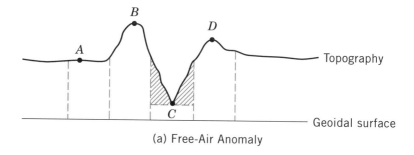

(a) Free-Air Anomaly

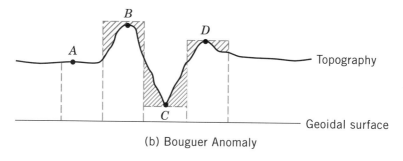

(b) Bouguer Anomaly

50. Gravity Anomalies

Free-Air Anomaly

The *free-air anomaly* reduces the observed value to the geoidal surface on the assumption that there is no mass between the earth's surface and the geoidal surface. This height correction is 0.3086 mgal for every meter above sea level. In other words, the point of observation is considered to be in free air above the geoidal surface, and while the height is taken into account, the intervening mass is not. However, where there is a mass above the observation point that will affect the reading, (such as for point C in figure 50a), then the free-air anomaly does not account for this effect. Free-air anomalies correlate positively to the topography and are assumed to be balanced by the compensation for depth. Except in mountainous areas, this is a reasonably accurate method.

Bouguer Anomaly

This is an extension of the free-air anomaly with an allowance for the topography between the two surfaces. The *Bouguer anomaly* assumes a constant density value for that mass and that the area around the observation point is a level plane. Thus, if the point is on a sharp peak (such as B in figure 50b) correction is made for a nonexistent mass surrounding it, and a positive correction is needed. The converse applies to a point in a valley (such as C in 50b), where the observed value has been overdiminished and needs a positive correction. An approximate value of this correction is 0.1119 mgal for every meter of elevation.

This correction, together with the application of the free-air reduction, gives the Bouguer reduction. What this does not allow for is the nearby topography.

Corrections for the effect of topography are made to a radius of at least 166.6 km from the point of observation. This apparently strange value arises from the sizes given to a series of zones and compartments around a point, as devised by John Hayford in 1907 to aid the reduction of gravity observations. It equates to 100 miles, the value Hayford no doubt initially used. The mean elevation of each compartment is used in the topographic reduction. For ocean areas, any measured values have to be corrected up to sea level according to the depth at which they were recorded—usually the depth at which a submarine took the instruments.

These anomalies are extensively used in geophysical prospecting. Over wide areas they are closely related to height and are generally strongly negative for mountain stations.

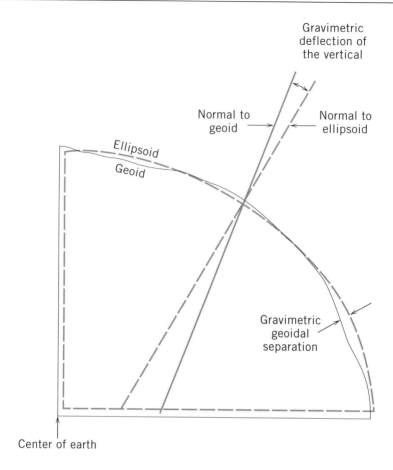

Gravimetric
deflection of
the vertical

Normal to
geoid

Normal to
ellipsoid

Ellipsoid

Geoid

Gravimetric
geoidal
separation

Center of earth

51. Products of Gravimetric Method

Isostatic Anomalies

These can be determined according to a particular hypothesis. In addition to the hypotheses of Pratt and Airy is a variation of the Pratt hypothesis developed by Hayford early in this century. Other scientists, such as William Bowie, Friedrich Helmert, and Weiko Heiskanen, contributed similar ones.

The theoretical value at sea level is corrected for elevation, for the total global terrain effect of topography, and for compensation according to the chosen hypothesis. Application of the total global terrain effect to the Bouguer anomaly gives the isostatic anomaly according to the chosen hypothesis. The basic variation between the various hypotheses is the chosen depth of compensation, usually taken to be either 96 km (60 miles) or 113.7 km (about 70 miles). Best results are obtained by varying the compensation depths regionally.

There are two schools of thought on the reduction of gravity observations. One approach adjusts the observed values and compares the results with the theoretical values (usually by countries outside the United States), while the other does the reverse, adjusting the theoretical values for comparison with the observed values (particularly in the United States). Obviously both methods should produce identical results.

GRAVIMETRIC METHOD

The method providing the basis for determining the undulations of the geoid from gravity data was published in 1849 by the British scientist Sir George Stokes (1819–1903). However, the lack of sufficient observed gravity data prevented its application for 100 years. In 1928 the Dutch scientist Felix Vening Meinesz (1887–1966) developed the work of Stokes into formulae by which the gravimetric deflection of the vertical could be computed (figure 51).

Computing the undulations of the geoid and the deflections of the vertical requires extensive gravity observations. The areas immediately surrounding the computation point require a dense coverage of gravity observations, and detailed data must be obtained out to a distance of about 800 km (500 miles). A less dense network is required for the remaining portion of the earth. While the observational requirements for these computations appear enormous, computer facilities now available relieve much of the tedium, and the results well justify the necessary survey work.

The observed elements of the *gravimetric method* are the gravity anomalies—the differences between the observed gravity value properly reduced to sea level, and the theoretical gravity value obtained from the international gravity formula. The gravity anomalies are caused by either local or regional mass variations, which may be visible or invisible, topographic or subsurface.

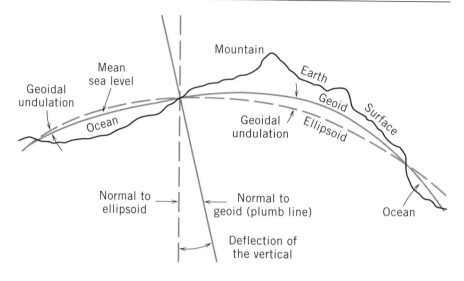

52. Geoid-Ellipsoid Relationships

Figure 52 shows how the mass surplus of the mountains and the mass deficiency of the oceans cause the deflection of the vertical and the undulations of the geoid. A mountain mass appears to "pull" the plumb line from the normal of the ellipsoid. In a similar way, the mass deficiency of the oceans appear to "push" the plumb line. These effects of the mass anomalies contribute to the deflection of the vertical. However, both deflections of the vertical and undulation values result from density variations throughout the earth. In the area of mass surplus, the observed gravity (reduced to sea level by the elevation correction only) is generally greater than the theoretical values, so the anomalies are positive. In the areas of mass deficiency, the observed gravity, reduced in the same way, is generally smaller than the theoretical value, so the anomalies are negative.

The deflections and undulations computed with sufficient gravity information are considered absolute values referred to an earth-centered ellipsoid. In other words, the axis of rotation for the ellipsoid coincides with the earth's center of gravity.

Effective use of the gravimetric method is dependent only on the availability of anomalies in sufficient quantity to achieve the accuracy desired. Successful use of Stokes's integral and Vening Meinesz's formulae depends on a good knowledge of gravity anomalies in the immediate vicinity of the point under consideration and a general knowledge of anomalies for the entire earth.

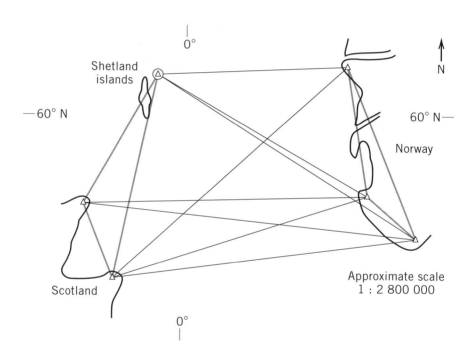

53. Shoran Connection between Scotland and Norway

CHAPTER
7
World Geodetic System

Because of the failure of local or even preferred systems to provide adequate intercontinental geodetic information, a unified world system became essential. To develop such a system, all of the military services actively participated in a program to develop a U.S. Department of Defense World Geodetic System (DOD WGS 60). To establish a world system, it was necessary to consider all the available observed data to determine the absolute reference system that best provided a good fit for the entire earth. Once such a system was established, all the major geodetic networks of the world could be unified and the coordinates of points anywhere on the earth would be compatible.

DERIVATION OF WORLD GEODETIC SYSTEMS

The development of a world geodetic system, however, would require enormous effort, particularly in the following areas:

1. establishing intercontinental links
2. measuring as much gravity data as possible
3. deriving approximate gravity values in blank areas
4. defining a suitable ellipsoid
5. deriving the gravimetric orientation

Let us consider each of these areas in turn.

Intercontinental Links

Because of its intercontinental-range missile capability, the U.S. Air Force played an active role with other agencies over a 10-year period to gather sufficient data to formulate a unified world geodetic system. Of particular use in this work was applying the results from Shoran and Hiran.

Shoran (Short Range Navigation) was a system for measuring distances to the required target from either a vehicle or fixed ground stations. Other types of system had the receiver at the required point and range information dictated from fixed ground stations.

Shoran was developed in the United States for use in World War II, but in 1944 was experimented with for civilian survey use under the guidance of Carl Aslakson at the USCGS (U.S. Coast and Geodetic Survey). Experiments consisted of flying at high-altitude-3000 + m (10 000 + ft)—across the line joining two stations. Ranges recorded to each station during repeated crossings gave results with a standard error for the total range equivalent to about 1 in 100 000, although in absolute terms for the distance, this was considerably downgraded by inaccuracies in the adopted velocity of light. (At that time Bergstrand was only just beginning his work on the velocity of light.)

The technique showed promise, particularly in the development of large trilateration networks in remote areas and of long-distance ties such as between Norway and Scotland (figure 53) and from Crete to North Africa.

Hiran ("high-precision Shoran") was developed from Shoran in 1949 to give higher precision. The basic mode of operation was similar to Shoran, although Hiran incorporated improvements in the electronics. Tests over similar long lines using a better value for the velocity of light gave standard errors in the distance of up to 1 in 100 000.

The Hiran tie from Canada to Norway permitted connection of the North American and European datums. However, since both of these datums are based on ellipsoids, which provide a good fit only in the areas of their origins, connecting the two datums through the Hiran tie overextended the usefulness of either ellipsoid. Nevertheless, ties made with Hiran provided an effective check on the two major datums oriented to a common ellipsoid. Other Hiran trilateration loops were valuable in the evaluation and extension of the DOD WGS 60.

As illustrated in figure 53, while Shoran and Hiran were forms of trilateration, the geometry of the networks was necessarily different from a normal land-based trilateration (see figure 27).

Collection of Gravity Data

A major area in which preparatory work was done in anticipation of a world geodetic system was in the collection and analysis of gravity observations. Through an extensive program several thousand gravity base-reference stations were interconnected throughout the world.

The connection of these base stations by the easily transported gravimeter permitted the reduction of numerous gravity observations to a common usable system. In effect, it established a world gravity system by including data collected for oil prospecting and other geophysical purposes.

The collection of extensive gravity data facilitated preparation of mean gravity anomalies for $1° \times 1°$ and $5° \times 5°$ geographic sectors as needed for further gravimetric computations. It also provided data for the preparation of the gravimetric geoid used for orienting the ellipsoid for the Army Map Service World Geodetic System (see page 135). Much of the work in the collection and analysis of the gravity data was done by contract with universities.

Gravity Values in Blank Areas

An additional phase of gravity work involved computing gravity anomalies for places where there were no observations. As stated before, gravimetric computations require a knowledge of gravity anomalies over the entire earth. Some regions of the earth were completely void of observation points. For these the best substitute was found by considering the topography or bathymetry (underwater topography) near the required positions and making certain applications of knowledge of the earth's crust. Although the computations for each position were tedious and involved consideration of the entire earth, the approximate anomalies were better than blank observation ''squares'' on the gravity anomaly map.

However, filling blank areas with approximate anomalies was not a substitute for actual observations when the latter were possible. Thus work continues gathering data by land, oceanic, and airborne (including satellite) gravity surveys.

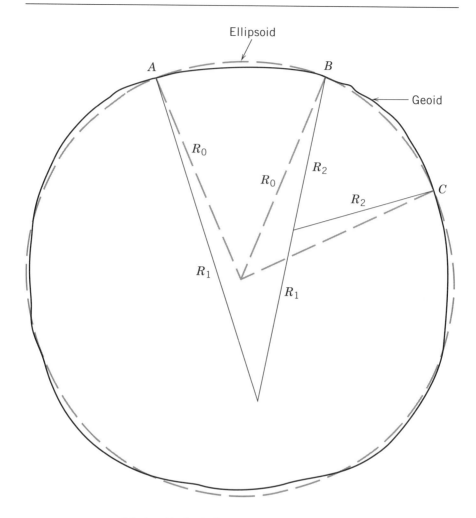

54. Arc Method of Determining Earth's Size

Derivation of Ellipsoid

Determination of the size and shape of the ellipsoid of revolution that provides a close approximation to the true figure of the earth is essential to the establishment of a world geodetic system. There are several ways in which this information can be obtained. The historical synopsis in Chapter 1 described the methods by which the ancient Greeks and others obtained their ideas of the size of the earth.

The method of arc measurement has not been changed in principle to the present day, although certain important refinements have been made. Basically, a long arc was measured on the earth's surface, and by means of astronomical observations the angle subtended by the measured arc was determined from the difference in latitude or longitude. (Although most arcs lay along a meridian, some were deliberately set along a line of latitude). By using appropriate formulae it was possible to compute the value of the semimajor axis and the flattening. From this procedure it was but a simple step to combine all available arcs into one calculation so that the result represented a wider area of the earth.

It has already been indicated that astronomical observations are based on the geoid. This causes errors in arc measurements that must be eliminated in order to obtain accurate results. Figure 54 illustrates how the deflection of the vertical can cause an error in computing the radius of the arc. The arc AB, where the geoidal surface lies below the reference ellipsoid, gives too large a radius R_1. The arc BC, where the geoidal surface is above the ellipsoid, gives too small a radius R_2. This error can be reduced by considering the effect of topography and the structure of the earth's crust surrounding the points where the astronomical observations are made. Still better results can be obtained if the deflections are computed by the gravimetric method. Removing the deflection of the vertical at points A, B, and C eliminates the error, so that the two arcs AB and BC give the real radius R_0 of the ellipsoid.

The ellipsoid used for the world geodetic system was based on all available information. The data reduced from arc measurements in the Eastern and Western Hemispheres as well as the connecting tie resulting from the North Atlantic Hiran trilateration arc were all considered. The arcs provided information for determining the size of the ellipsoid.

Artificial earth satellites furnished observational data for determining the earth's flattening. For example, the 1958 *Vanguard* satellite had been tracked several thousands of times, and data reduced from observations indicated a probable flattening factor of 1/298. This value had been suggested earlier by Friedrich Helmert and also by Theodosy Krassovsky, who used it as an element of the 1942 Pulkovo datum. More recently, satellites designed specifically to collect relevant data have modified the accepted value just slightly, to 1/298.26.

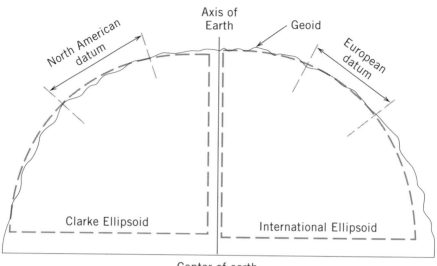

55. Preferred Datums Are Relative Datums

Gravimetric Orientation

The preferred datums (figure 55) were given a relative orientation with respect to a certain portion of the geoid by the methods already outlined (see figure 54). To convert each of them into a single earth-centered world system, it was necessary to change the size and shape of each ellipsoid and reorient them so that the center of the new ellipsoid would coincide with the center of the earth and the semimajor axis would coincide with the earth's axis of rotation. This reorientation was achieved by the gravimetric method.

To obtain an absolute gravimetric orientation for the preferred datums the following procedure was necessary:

Compute the gravimetric geoidal height and deflection of the vertical for the initial point of the geodetic system Theoretically the absolute undulation and deflection components for one astrogeodetic point in each geodetic datum would be sufficient to reduce the datums to a world system. However, by using the mean of the gravimetric quantities of several stations around a selected point the inaccuracies likely to result from error in the astronomical coordinates or gravity information around a single point are minimized. For this reason, the gravimetric components for deflection of the vertical and geoidal undulation were computed for a group of astrogeodetic stations in the vicinity of the origin of the datum, and the astrogeodetic minus gravimetric differences were transferred to the origin by transformation formulae. The results were further improved by repeating the procedure for a second area within the same datum and taking a weighted mean of the results for the two areas.

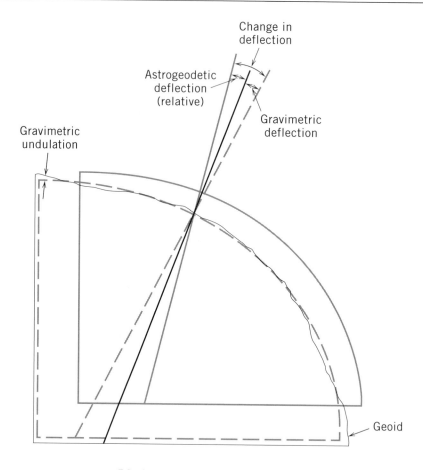

56. Gravimetric Datum Orientation

As a correction, apply the mean of the differences between gravimetric and astrogeodetic deflection components and geoidal undulation at the initial point See figure 56.

Use transformation formulae such as those devised by Vening Meinesz or J. de Graaff-Hunter to compute the correction for each station in the preferred datum This procedure was applied to the North American, European and Tokyo datums to establish the USAF World Geodetic System 1958. Later revised to incorporate additional survey and satellite information, it became the USAF World Geodetic System 1959.

Deflection of the vertical components were computed for each of some 30 points by the Vening Meinesz procedure and geoidal heights were determined according to Stokes's formulae.

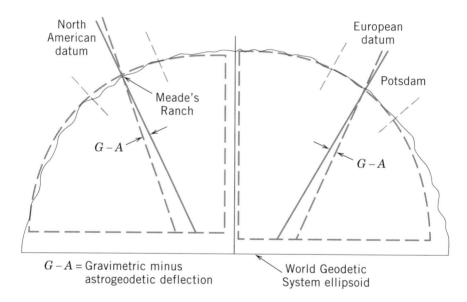

North American datum

European datum

Meade's Ranch

Potsdam

$G - A$

$G - A$

$G - A$ = Gravimetric minus astrogeodetic deflection

World Geodetic System ellipsoid

57. Deflections of Vertical Computed from Earth's Gravity Can Determine a World Geodetic System

ARMY MAP SERVICE WORLD GEODETIC SYSTEM

Shortly after the USAF World Geodetic System of 1958 became operational, the U.S. Army Map Service (AMS) published its results of work on the development of a world system. Its fundamental concepts were consistent with those of the Air Force.

The Army studies relied largely on geoidal heights for determination of the best-fitting ellipsoid and for obtaining the necessary earth-centered orientation of the ellipsoid. Using deflections at numerous astrogeodetic stations, astrogeodetic geoidal undulations were computed for the North American datum in North and South America and for the European datum in Africa and the Eurasian continent (figure 57). These were compared with gravimetrically computed undulations, and an adjustment was used to minimize the differences between the astrogeodetic and gravimetric geoids.

By matching the relative astrogeodetic geoids of the preferred datums with an earth-centered gravimetric geoid, the preferred datums were reduced to an earth-centered orientation, thereby producing the U.S. Army's tentative world datum of 1958. In 1959 after additional information—Hiran data and a better flattening factor—was incorporated, the improved system became the U.S. Army World Geodetic System 1959.

U.S. DEPARTMENT OF DEFENSE WORLD GEODETIC SYSTEM 1960 (DOD WGS 60)

The Aeronautical Chart and Information Center (ACIC) and AMS had each developed a world geodetic system by different procedures. ACIC derived the USAF World Geodetic System 1959 using differences between the astrogeodetic and gravimetric deflections at specifically selected stations in the areas of the major datums. AMS derived the U.S. Army World Geodetic System 1959 using a comparison of astrogeodetic and gravimetric geoidal heights in the areas of the major triangulation systems. Both systems incorporated satellite observational data and intercontinental Hiran trilateration information.

Since the two systems agreed remarkably well for the North American, European, and Tokyo datum areas, they were consolidated into a single system called the Department of Defense World Geodetic System 1960 (WGS 60). The WGS 60 provided a single, unified, earth-centered system through which accurate distance and direction information could be determined for missiles of all ranges.

LATER WORLD GEODETIC SYSTEMS

WGS 60 served well for some 12 years or so, but the increasing use of satellites drove the need for an even better system.

All processing of the global positioning system (GPS) observations was then performed on WGS 72, which was an earth-centered and earth-fixed system. The final WGS 72 latitudes and longitudes could be converted back to a specified datum, but there were problems with heights unless the geoid-ellipsoid separations were known for that area.

There is also a WGS 72 ellipsoid, where the semimajor axis $a = 6\,378\,135$ m and $1/f = 298.26$.

Yet another new system—WGS 84—has now been adopted and is taking over the previous role of WGS 72. WGS 84 is almost identical with NAD 83. Although they are both defined as geocentric, they are, according to latest information, a meter or two away from being exactly so.

OTHER NOTABLE SYSTEMS

New systems and new abbreviations appear almost overnight, and it is quite impossible, and even unnecessary in some instances, to keep up with all of them. Several pages could be devoted to listing and explaining the abbreviations, but just a few of the more important ones will suffice here.

EUREF

A subcommission of the International Association of Geodesy (IAG) was responsible for defining the European Reference Frame (EUREF), a true tridimensional control network covering Europe. The frame was defined using very high accuracy techniques such as very long baseline interferometry (VLBI) where a few centimeters can be achieved over thousands of kilometers.

ETRF 89

The European Terrestrial Reference Frame 1989. (ETRF 89) resulted from EUREF and aimed to give more precise coordinates throughout Europe. It is compatible with WGS 84 to within 1 m and is the fundamental geodetic reference frame in Europe. It is not directly comparable to either the OSGB 36 (Ordnance Survey Great Britain 1936) or its equivalents in other countries, and the differences are not uniform.

ETRF 89 was used to transform the GPS coordinates in terms of WGS 84 to their equivalents on the Ordnance Survey Geodetic Reference System, 1980 (OSGRS 80) and thence to the national grid, and vice versa. The procedure, for which tables have been derived, is to convert the GPS values to plane coordinates (or OSGRS 80) on the Transverse Mercator projection. These values are then subject to a shift to get them to the national grid.

ETRF 89 is based on the GRS 80 spheroid. While it is now reasonably in sympathy with ITRF (see below) and WGS 84, tectonic movements over the continent are of the order of 2.5 mm yr, so that as time goes by, ETRF 89 will differ increasingly from ITRF.

ITRS and ITRF

The International Terrestrial Reference System (ITRS) has been adopted by the International Union for Geodesy and Geophysics (IUGG) for geodetic and geodynamic purposes. ITRS is maintained by the International Earth Rotation Service (IERS), which was founded in 1987 to embrace the work previously carried out by several other bodies. IERS has the task of defining and maintaining the ITRS through high-precision techniques such as VLBI, lunar laser ranging (LLR), and satellite laser ranging (SLR). The origin of the system is considered to be within 0.1 m of the earth's center of mass. In addition, IERS is active in relating tide gauges to International Terrestrial Reference Frame (ITRF) and carrying out a long-term monitoring exercise in three dimensions. The aim is to determine the variations in sea level without some of the disturbing parameters.

CHAPTER
8

Satellite Geodesy

With the rapid increase in industrial technology after World War II came the requirement for improved geodetic information. Major surveys and positioning techniques were still laborious and difficult, as well as dangerous, in the more inaccessible parts of the world. There was a need to integrate the surveys on a worldwide basis to relate all the major datums of different continents. Traditional survey techniques were not sufficient without the injection of a new approach. Rocket developments during the war provided the required stimulus.

In theory, had it been possible to position objects in predictable orbits around the earth, a considerable step would have been made in pushing the frontiers of geodesy forward. The breakthrough came in 1957 with the launch of the first artificial satellites, but hand in hand with this was the need for improved tracking methods and the development of high-speed computers. Without the tracking facilities, prediction of the orbital changes would be impossible yet if the changes were observable the amount of calculation required in short intervals of time would be unachievable. Thus it was essential for all three—satellites, tracking, and computing—to progress in parallel.

EARLY SATELLITES

The first *Sputnik* was launched from Russia on October 4, 1957. It weighed 184 lb (83 kg), was powered by ordinary chemical batteries, and emitted radio signals for just 3 weeks. With an average orbital height of 360 miles (576 km), it orbited in 96.2 minutes. The orbit was elliptical, varying from 136 miles (216 km) to 585 miles (1136 km), but air drag and gravity rapidly decreased these values so that it completed only 1400 orbits in a life span of 92 days. Although it had no experimental scientific instrumentation on board, it represented the forward step that science had been waiting for.

The second *Sputnik,* launched a month later, lasted for nearly 2400 orbits and had basic instrumentation for data collection.

America entered the space race in February 1958 with *Explorer I,* which was in a higher orbit than the *Sputniks,* varying from 222 miles (355 km) to 1593 miles (2551 km), and as a result had a life of 4 years.

From the geodetic point of view it was America's *Vanguard I*—launched in March 1958, with an even higher orbit, varying from 410 miles (656 km) to 2468 miles (3950 km), and estimated lifetime of 100 years—that made the most significant contribution.

It is interesting to note that this breakthrough for surveying was very much contemporary with another major advance—EDM. Although first used in the Geodimeter in the 1940s, toward the end of the 1950s such equipment began to be accepted as an everyday tool and additional models became available (see Chapter 14).

With the early satellites there were three main tracking systems—optical, radio, and radar. Optical techniques were affected by the size of the satellite as well as by adverse weather conditions. Radio methods had fewer problems but could not produce given accuracies as readily as the optical methods. Radar similarly had few problems, except that a satellite is a very small target at a considerable distance.

The simplest optical method of using the naked eye was not at all feasible and better results were possible with either an ordinary theodolite or a *kine-theodolite* (a theodolite with camera facilities). When a satellite was photographed against a background of stars it was theoretically possible to resolve the angular position to around 3″ of arc. This presented some difficulties, however, when it came to higher accuracies because of the size of the required camera, such as the Baker-Nunn camera, and the visibility problems of such an arrangement. Satellites designed specifically for optical observation contained a light source which could be activated electronically for photographic purposes. A large satellite with no independent source of illumination could be photographed at dawn or dusk when reflected sunlight shone on it. Additionally, the satellite needed to be bright enough both to allow its location, and to leave an image on the film. Radio methods could use either interferometry or Doppler techniques (see below).

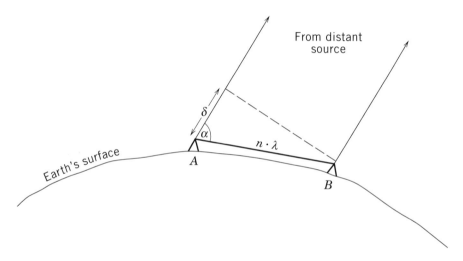

58. Principle of Radio Interferometry

INTERFEROMETRY

Interferometry (figure 58) required that two aerials be positioned at the same height and set at a distance apart equal to an integer number n of wavelength λ of a particular signal. When the two sets of received signals were in phase, then the difference in distance from the satellite to each aerial would also be an integer number of wavelengths. If the signals completely interfered, then the difference in distance was a multiple of half the wavelength. At long wavelengths there were problems with bending of the radio waves.

DOPPLER

The Doppler technique was first investigated by Christian Doppler in 1842. One of its early survey applications was its use in Antarctica during 1972; at that time it was particularly bulky, although accuracies of a handful of meters were possible.

The Doppler phenomenon is usually explained in terms of the movement of a noise source in relation to a stationary observer (figure 59). If the source moves in such a way that it remains at a constant distance from the observer (that is, in a circle), then the pitch will be constant because the wave always travels the same distance. If the source travels in a straight line, it must move toward and then away from the observer. As it does this, the waves have decreasing distances to travel; as they appear to "pile up," the detected pitch or frequency increases. As the source moves away from the observer, the reverse effect becomes evident.

If a second observer is farther away from the source, then the change in pitch would be more gradual. If the sounds at the observer position were recorded and analyzed, it would be possible to determine how far the source was from the observer. Note that an example like a train involves sound waves, while a satellite involves radio waves. The effect is the same, but radio waves travel at about 300 000 km/sec, while sound travels at only 330 m/sec. The basic relationship is that

$$\text{change in frequency} = \text{frequency} \times V/C$$

where

$$V = \text{velocity of source}$$
$$C = \text{velocity of radio waves}$$

Radar methods sent short wavelength pulses (1 mm to 10 m) to the satellite and endeavored to monitor the amount of reflected energy. From this, use of signal travel time and the velocity of light gave the distance. For

S_3 = point of nearest approach
$S_2 - S_3$ = signal increases pitch
$S_3 - S_4$ = signal decreases pitch

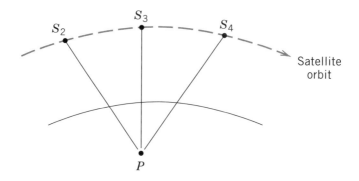

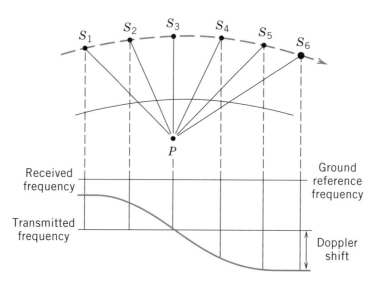

59. Doppler principle

accurate results the beam had to be narrow, which, in its turn, made it more difficult to hit the target.

Note that the velocity of light and the velocity of radio waves are the same when in a vacuum. In the atmosphere they vary slightly due to different effects of pressure, temperature, and humidity, but for the purposes of this text, we can consider them equal at about 300 000 km/sec.

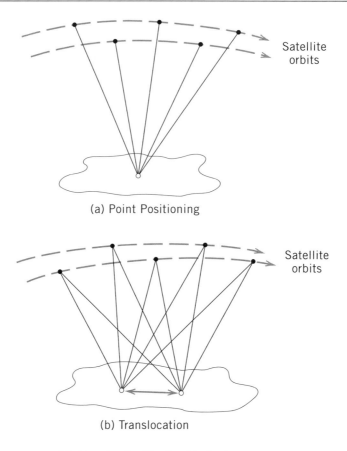

(a) Point Positioning

(b) Translocation

60. Doppler Positioning Methods

The *Doppler count* is the number of cycles of the *beat frequency* (the difference between the frequency generated at the receiver and that received there from the satellite). There are three ways of determining the Doppler count: over discrete time intervals, as a cumulative total over a full passage, or in terms of the time taken to receive a given number of cycles. This last method is not suitable for field survey work. If the count is combined with knowledge of the satellite orbit parameters, then the position of the receiver can be found with respect to the earth's center of mass—or geocentric coordinates.

Doppler observations can be reduced by at least three different methods: (1) point positioning; (2) translocation, and (3) short-arc translocation.

Point Positioning

The process *point positioning* (figure 60a) assumes that the ephemerides (set of orbital parameters) of each pass are correct. Data is collected for a number of passes at a single station. The ephemerides may be either "precise" or "broadcast." Precise ephemerides are derived from observations over a 48-hour period from a worldwide network of over 20 monitoring stations. Accuracy of the data is 2 to 3 m. Broadcast ephemerides are encoded onto the satellite signal so that they are immediately accessible. The data come from tracking results at four stations only, and hence accuracy is only some 20 to 30 m. Broadcast ephemerides are essentially a forecast of the satellite position, whereas the precise ephemerides are derived from observations on the actual orbit used. For point positioning the data reduction is relatively simple but is of a low order of accuracy.

Translocation

Translocation (figure 60b) is a modification of point positioning, in which two points are used. Passes are recorded simultaneously at each station. If the points are not too far apart, some sources of error, particularly in the ephemerides, can be considered equal and canceling out. This allows a far better accuracy in the relative positioning. An extension of this is to keep one receiver stationary while a second one occupies, in succession, a number of required points—rather along the lines of the single-base method of barometric heighting.

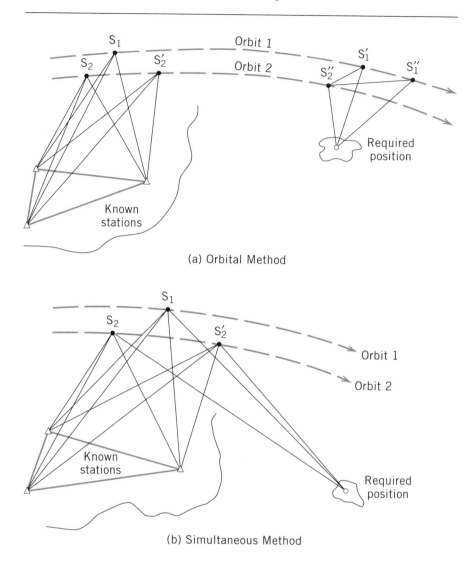

(a) Orbital Method

(b) Simultaneous Method

61. Simultaneous and Orbital Methods First Used with SECOR

Short-Arc Translocation

Short-arc translocation is a variation of the orbital method (figures 61a and 62). It uses a portion of the satellite arc. The network allows a rigorous adjustment procedure to be applied, but the computational requirement is considerable since it allows adjustment of the orbit as well as the ground coordinates. It can be modified in various ways to simplify the procedure.

By extrapolation from the results of the short-arc observations, the orbit is defined for a far distant station that observes the same satellite pass. From a set of at least three known positions, simultaneous observations (figure 61b) to the satellite allow definition of the orbit. Stations at unknown points, distant from each other so as to be unable to observe the satellites at the same time, make separate observations. The desired ground positions are then determined by a form of three-dimensional resection from the satellite positions.

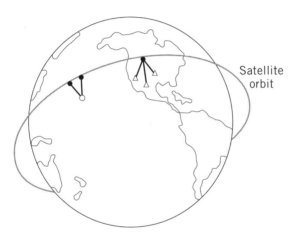

Satellite orbit

△ = Known positions
○ = Required position

62. Short-Arc Method

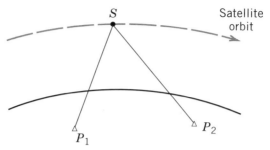

(a) One Satellite, Two Receivers:
Cancels Satellite Clock Offset.

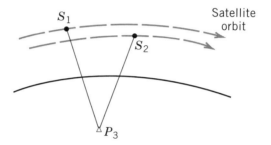

(b) Two Satellites, One Receiver:
Cancels Receiver Clock Offset.

63. Single Difference

TRANSIT DOPPLER

Transit Doppler is a U.S. Navy Navigation Satellite System. Under development since 1958, it has been in existence since 1964 and available for civilian use since 1967. It is expected to continue at least until the late 1990s when GPS should be fully phased in. By the end of 1981 there were six satellites in orbit for use by this method and 12 more in store.

They weigh only 61 kg, are in near-circular polar orbits just over 1000 km high, and orbit about the earth every 107 minutes. From a sequence of range measures during 10 minutes a good fix can be obtained in various configurations (figures 63 and 64) for either a moving vessel or a stationary ground station. Point positioning is good to 1 to 5 m depending on whether precise or broadcast ephemerides are used.

It has been proved to be a very reliable system; users of this and later satellite systems appear to be growing by some 50% each year and is approaching the tens of thousands. With the small number of satellites there could be long (6- to 12-hour) waiting periods between available fix times. Today there are only four satellites transmitting. When there were seven in operation, an acceptable three-dimensional pass could be obtained in Great Britain almost every hour, but with the reduction to four that is not now possible.

The satellites transmit at 150 and 400 MHz with very stable frequencies. The phase-modulated signals include time information as well as orbital data. The position of the receiver in relation to the satellite is found by the Doppler shift described on page 141. Each unit of the count is less than 1 m, so it is a very sensitive system.

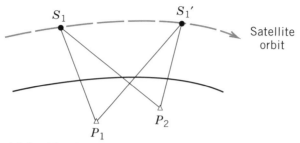

(a) Double Difference: Both clock offsets cancel.

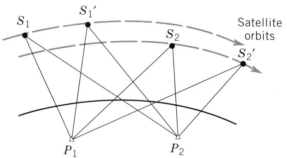

(b) Triple difference: Both clock offsets cancel; phase ambiguity is eliminated.

64. Double and Triple Difference

Any convenient satellite pass is suitable for observation if it is above the horizon in a more or less overhead orbit during some 18 minutes, which is time enough for up to 40 counts.

Refraction due to the ionosphere causes the satellite to appear to be on a path of greater than actual curvature, which tends to reduce the Doppler shift by "moving" the position of the receiver up to 0.5 km. The troposphere also produces refraction errors by affecting the propagation velocity of the signal. It is not nearly so easily detected as the ionospheric effect. The dual frequency helps to minimize errors due to refraction.

Results of transit Doppler observations are on the world geodetic system (WGS 72) and can be subsequently transformed to any required national datum. This can "change" the geographic positions of a station by the equivalent of several hundred meters.

For a relatively small outlay on a receiver, transit Doppler provides an all-weather, high-accuracy positioning system. It has found a wide range of applications including pipeline positioning, oil-well location, determining boundaries of all types as well as isolated points throughout the world as parts of national survey schemes.

Accuracies with this system can be down to 0.25 m with observing periods from a half day up to 2 days. The system has four transmitting satellites all in polar orbits of about 1100 km, which gives them orbit periods of about 107 minutes.

OBSERVATIONAL SYSTEMS

Initially two basic systems were used for obtaining geodetic information from artificial earth satellites through three principal observational methods. Optical and electronic facilities made it possible to perform various geodetic measurements by simultaneous, orbital, or short-arc methods to determine geocentric position, relate unknown positions to established points, and extend existing triangulation networks into unsurveyed areas.

Among the early developments in satellite observation was SECOR (SEquential Collation Of Range). This consisted of a transponder that received, amplified, and retransmitted high-frequency signals from ground stations. While it was an all-weather system and was described as mobile, it did need a large truck to transport it—by today's standards it was very large, cumbersome, and heavy. Yet it allowed determination of the coordinates of positions up to 2500 km (1500 miles) from known points.

SECOR was developed by a contractor to the U.S. Army Corp of Engineers. The first satellite used with it was launched in 1964. To avoid uncertainties in the earth's gravity field, SECOR was used in a geometric mode only. This mode used a total of four ground positions—one of which had to be designated as master—from which simultaneous range observations were made to a transponder in the satellite.

Of the four positions, only one could be unknown, then; rather along the lines of resection, a set of range values allowed computation of the unknown position. In practice, of course, many sets of observations were taken to allow a least-squares solution. Such an operation required the three known stations to provide a well-conditioned triangle. The observations were best when the satellite was above the center of that triangle.

The two main sources of error were in the system calibration and in the effects of refraction in both the troposphere and ionosphere. Tests suggested that the achievable accuracy was around 3 ppm probable error with improvements from night observations.

In practice, the sequence of operations required to complete a positioning was repeated every 50 milliseconds. It required a minimum of two satellite passes. For better results, data from positions on both passes (or more if possible) were combined to give improved geometry. With four ground stations for SECOR it was possible to measure some 29 000 ranges during a normal 6-minute satellite pass—a position fix every 1/20 second. All data were recorded on seven-channel magnetic tape and then sent to a central processing agency.

In fact, the fixes did not occur exactly every 1/20 second but only on average: when the satellite was close to the stations, it could be less than 1/20; conversely, when farther away, the time could be slightly longer.

The coordinates of the unknown point will be in the same coordinate system as the known points. Such a system of control points can be readily extended.

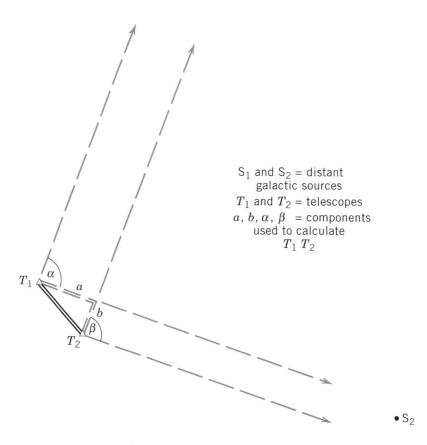

S_1 and S_2 = distant galactic sources
T_1 and T_2 = telescopes
a, b, α, β = components used to calculate $T_1\,T_2$

65. VLBI Baseline Determination

CHAPTER
9

Very Long Baseline Interferometry

In the second half of the 1960s astronomers developed a technique, called *very long baseline interferometry* to improve the resolution of their radio telescopes based on extraterrestrial objects such as quasars. Geodesists thought the idea might be applicable to geodesy and geophysics as well as astronomy. By the early 1970s geodetic experts in the United States were experimenting with it.

The concept is that a radio signal from a quasar or similar source is received at two or more distantly separated parabolic radio telescopes that form an interferometer. The separation of the telescopes means the signal will have different distances to travel. The accurate times of arrival of the signal can be processed, in conjunction with the geometry of the situation, into information relating to the relative positions of the telescopes.

The difference in recorded time of arrival of a signal at the two telescopes—or relative phase delay—is directly related to the component of the baseline in the direction of the source (figure 65). Thus rotating the telescopes and using a series of sources more or less paired at right angles to each other will theoretically allow calculation of both the length and direction of the baseline. The signals are recorded on tape and processed at a central location rather than locally.

This description somewhat oversimplifies the process since the earth is moving in relation to the distant sources. Allowance has to be made for this effect together with other factors such as refraction, precession, nutation, and offset of the clocks.

Obviously as time differences are used it is essential to know the relationship between the two clocks. The time standard can be based on rubidium or cesium at stabilities of at least 1 in 10^{11}, or on the hydrogen maser at 1 in 10^{13} over a period of 1000 seconds (16 to 17 minutes). The refraction effect can be around 10^{-8} seconds, or 2.5 m in distance.

Accuracy for the resulting separation of the telescopes is more or less independent of length and can fall in the centimeter region even over ranges reckoned in thousands of kilometers. Orientation is in parts of a second of arc. As with so many other aspects of surveying, it is the atmosphere that is the limiting factor.

In addition to determining the relative location of the telescopes, VLBI is also the best technique for determining the earth's rotation parameters, plate motion, and crustal movements. VLBI also finds application in linking survey networks of far-flung countries in different continents, and in establishing a global geodetic reference system.

The terminals of such an arrangement are usually the electrical centers of the antennae, so for survey purposes, local connections by traditional methods are required to connect to the nearest part of the national survey network.

VLBI and laser ranging are closely related. They can supplement one another when there is a requirement for a high-precision goedetic network as for some monitoring schemes. The most meaningful use of VLBI, in the geodetic sense, is when at least four receiving stations are observing simultaneously. This allows the separation of some of the parameters and the solution of others.

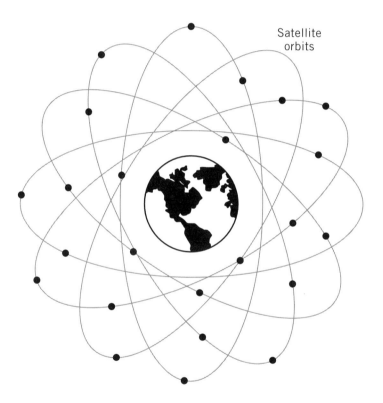

Satellite
orbits

66. Global Positioning System's Orbital Configuration

Global Positioning System

BASIC DESCRIPTION

The modern approach to the use of satellites for positioning is the NAVSTAR (Navigation System using Time And Ranging), more usually know as the *global positioning system* (*GPS*). Such are the expectations of this system that within just a few years it has replaced much of the traditional survey work in both the horizontal and vertical planes. The prime obstacle to using it has been the cost, which started at many thousands of dollars or pounds per unit, but which is now around £10 000 ($16 000) for two single-frequency receivers. To make it really effective and widely used, expenditure for the equipment should decrease to no more than that for a good theodolite. In addition, the processing must all be on the spot and not require dispatch to any central location. Its aim is to allow virtually instantaneous determination of position to a relatively low accuracy or, by occupying more observing time, a high-accuracy positioning in three dimensions.

GPS is designed in such a way that, now it is fully operational, it allows continuous provision of information worldwide. It is an American development conceived in the 1970s and for which 11 prototype satellites were put into orbit between 1978 and 1985. Two generations of GPS satellites have been developed, and a further two are scheduled for launching by the year 2000.

The satellites are generally launched from a space shuttle, and the *Challenger* disaster of 1986 was a serious setback to GPS. It had been hoped that the full requirement would be completed by 1989, but this was delayed by several years. The full constellation of 24 satellites was in operation by 1994.

The full system consists of 21 active satellites and three spare ones in six different orbits (figure 66) at 55° to the equator, at heights of 20 200 km (12 620 ml), so that orbital times are 12 sidereal hours. Power comes from solar batteries. When in high orbits in relation to the observer (that is, passing nearly overhead), the satellites will be available for 4 to 5 hours before disappearing over the horizon. The arrangement of the orbits and positions of the

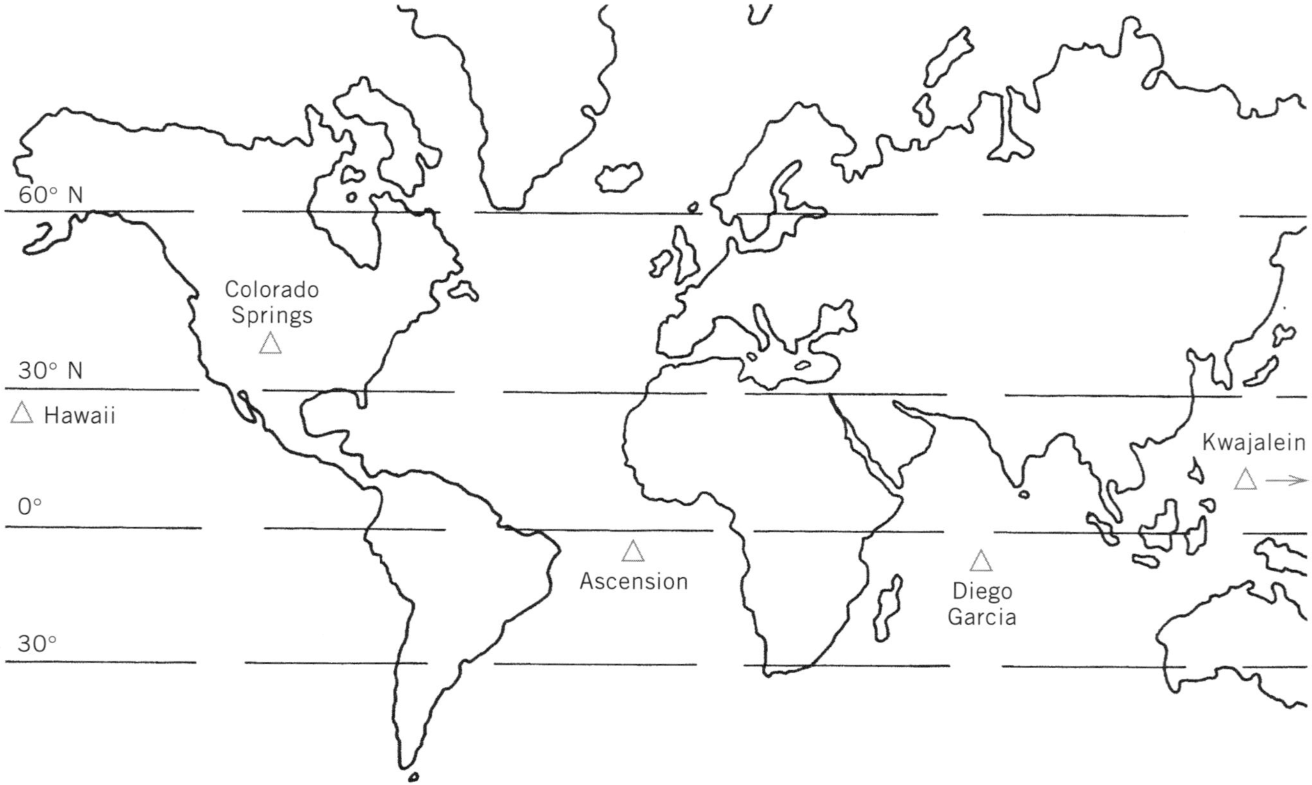

67. GPS Control Stations

Colorado Springs △

△ Hawaii

Kwajalein ▽ →

Diego Garcia ▽

Ascension ▽

60° N

30° N

0°

30°

satellites is such that at any time of the day or night at least four should be observable on an almost worldwide basis, although there are scattered areas in the latitude regions of 30° to 50° N and 30° to 50° S that have degraded coverage once or twice a day.

The control of the system is maintained from five stations spaced around the world. The master station is at Colorado Springs (39° N 105° W), although it was formerly at Vandenberg, California, the others are at Ascension (8° S 14° W); Diego Garcia (10° S 75° E), Hawaii (20° N 155° W), and Kwajalein (9° N 167° E) (figure 67). These five stations lie on four different major tectonic plates; the two major plates not represented are those for Africa and Eurasia.

Essential requirements for GPS are, as for VLBI, accurate frequency standards and clocks. These can be rubidium, cesium, quartz crystal, or hydrogen maser. In addition, good ephemerides—catalogues of predicted positions of satellites against time coded into satellite signals—are necessary. Just as errors in the clock or in the range can affect the result, so can inaccuracies in the orbital predictions. Any ephemeris error assumes the satellite to be at an erroneous position, and this affects the range. A satellite moves some 3 km/sec when in a 12-hour orbit, so even a small time error or prediction error can mean appreciable ground position error.

OTHER POSITIONING SYSTEMS

A USSR system similar to GPS is *Glonass,* which also has 24 satellites but in three orbits. Glonass started in 1982 and reached full operation on January 18, 1996. Unlike GPS, this is a purely civilian system, and there are receivers available capable of accepting both GPS and Glonass. When these dual receivers become fully reliable, they will have the advantage of allowing observations to more satellites at any one time than each system does on its own.

Although both systems have the same number of operational satellites, the sets are at different orbital inclinations and heights. Glonass satellites are at a 64.8° inclination, a height of 19 133 km (11 960 miles), and orbit some 43 minutes quicker than the GPS ones. Glonass transmits in the L band frequencies of 1597 to 1617 KHz and 1240 to 1260 MHz and uses either Moscow time or UTC. Its coordinates are based on the Soviet Geocentric Coordinate System 1985.

Other systems should be mentioned as an indication of the multiplicity of orbiting bodies that are in operation. Mention has already been made on page 149 of the Transit network. A USSR system similar to this is called Tsicada.

The European Space Agency (ESA) has developed Navsat (Navigation Satellite). Starfix is a geostationary complex of four satellites over the Gulf of Mexico operating in a geosponder mode only. That is, they receive signals from earth-based stations and retransmit them to other users. The United States has a proposed geostationary system, Geostar, which has yet to get past its development stage.

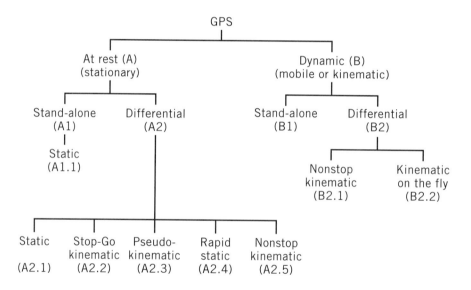

68. GPS Operating Modes and Techniques

GPS OPERATING MODES AND METHODS

GPS is a passive, all-weather system, available 24 hours a day. It can be used in either a stationary (at rest) or mobile (dynamic) mode, and used for either point positioning (stand alone) or relative positioning (differential) (figure 68).

A Stationary (at Rest) Positioning

In this mode, observations are made using a stationary receiver. Successive observations at a point can be to different satellites and at different times. The period during which the receiver has to be stationary can range from a few seconds to many hours, or in exceptional circumstances, a day or two.

B Mobile or Kinematic (Dynamic) Positioning

In this mode the receiver is in a moving vehicle. The results are used, for example, as a navigation aid at a low level of accuracy of 5 to 300 m. However, if this mode is used in conjunction with a second, stationary unit (that is, in a differential mode), accuracies from 5 m down to 1 cm are possible. With a mobile receiver there need to be at least three simultaneous observations. The receiver is positioned continuously, and a full route description is obtained.

A1, B1 Point (Stand-Alone) Positioning

This method locates individual points three-dimensionally within an established coordinate system (see figure 60a). The system may be local, or it may be worldwide, based on geocentric coordinates so as to give absolute positioning. The accuracy achieved in such a situation, using only a single receiver, is 300 m or better, although the actual figure in any instance is not readily deduced.

A2, B2 Relative (Differential) Positioning

This mode involves locating one point in relation to another point (known as translocation) (see figure 60b). Because of this it is more likely to use a local coordinate system, although it may be appropriate to later transform the values into an alternative system. Transformation would also be necessary if the two points in question were themselves on different coordinate systems. Relative positioning is simpler than absolute positioning, but it requires simultaneous observations from each point. Either of the two arrangements can be extended into a network of points, and then more sophisticated adjustment is possible with redundant observations. Often known as Differential GPS (DGPS), accuracies of a few millimeters are possible.

A1.1, A2.1 Static Positioning

In this technique, the receiver is stationary for around an hour, but the time will vary according to various factors and whether it is being used in a stand-alone or differential mode. As with EDM, some receiver models have different ranges from others. Single-frequency receivers usually have an upper limit in this mode of 15 km while dual-frequency models can cover 15 km and more, with no particular limit. Of all the possible variations, static positioning gives the highest accuracies and allows observation of the longest lines, although these can take several days to achieve.

A2.2 Stop-Go Kinematic Positioning

This is a differential or relative technique. It is a system akin to barometric heighting with one field and one stationary instrument or the single-base method. One receiver stays at a known point recording continuously. The second receiver starts at another known position and then moves successively around the unknown points, keeping lock on the satellite signals as it goes.

A2.3 Pseudo-Kinematic Positioning

This technique is a modification of A2.2. As before, there is a stationary receiver at one known point, and the moving receiver starts at a second known point. However, the moving receiver switches off before going to the first unknown point and also when moving between other unknown points. The drawback to this method is that each point has to be reoccupied after about an hour.

A2.4 Rapid Static Positioning

This is the latest technique and improves on A2.3 in that no reoccupation is necessary. The time spent at each unknown point is 5 to 10 minutes. Rather akin to single-base barometric heighting, a single receiver stays at a reference position and tracks continuously. A second receiver visits all the required field positions, with the receiver switched off between positions, and each point is treated individually in the calculations. This approach is particularly useful over short (5 km) baselines.

A2.5, B2.1 Nonstop Kinematic Positioning

In this method, despite its name, the mobile antenna has to be stationary for about 5 minutes to determine the starting point. Then it can be moved, on any sort of vehicle or just manually, along the required route to obtain a continuous record. It is essential to have four or five satellites visible; any less and a new start has to be made.

B2.2 Kinematic On-the-Fly Positioning

This is the same as the previous method without the need for the initial stationary period. The initialization can be achieved on the move and the ambiquities resolved if five or more satellites are visible.

Seeded Rapid Static Positioning

This is a method pioneered by the University of Pretoria for use in subsidence monitoring where a good approximation to the position is known prior to the observation. If the position is known to within one or two times the 19-cm wavelength, then it allows a resolution of the whole number ambiguities (see below).

AMBIGUITIES AND THEIR RESOLUTION

Satellites are able to provide either pseudo-range measurements or carrier beat phase measurements. In *pseudo-range measurement* the information required is the signal frequency, the exact time of signal transmission, and the exact time of signal receipt. Then, in effect, one applies the relation distance = velocity × time. The prime difficulty with the pseudo-range method is synchronizing the clocks. In *carrier beat phase measurement* the phase of the received signal is compared with the phase when it was transmitted. The difference between these gives only the fine part of the reading. The integral number of wavelengths has to be determined by other techniques. The integral number is known as the *initial phase ambiguity.*

GPS is a one-way ranging system where the controlling transmitters and user receivers have separate clocks and hence the need to know their interrelation or offset. Any discrepancy in the synchronization will directly affect the range by an amount related to the velocity of light (that is, 300 000 km/sec). Thus 1 km is equivalent to 3.3 microseconds (3.3×10^{-6} seconds).

The accumulation of errors from such sources as these can amount to a centimeter or two on short lines, or about 1 in 10^6 (1 ppm) of the distance as the line becomes more than a few kilometers. Some experts use the rule of thumb that uncertainty in the baseline is given by the expression

$$\text{base length} \times \frac{\text{ephemeris uncertainty}}{\text{satellite altitude}}$$

A particular advantage of the relative positioning approach is that some errors, common to both stations or systems, can be eliminated, or considerably reduced, with resulting improved accuracy (see figures 63 and 64).

As with ground survey techniques, the configuration of the observations can have a bearing on the accuracy of the results. Poor ''strength of figure'' can be detrimental.

The operation of NAVSTAR is based on signals continuously transmitted on two carrier frequencies of 1575.42 MHz (equivalent to a wavelength of 19 cm) and 1227.60 MHz (equivalent to a wavelength of 24 cm) multiplied up from the high-stability oscillator on the satellite of 10.23 MHz.

CODES

GPS uses two code arrangements: P(Y), or precision code; and C/A, or coarse acquisition code, sometimes referred to as S or standard code. The frequency of 1575.42 MHz carries both the P and C/A codes, that of 1227.60 MHz only the P code. From these two codes stem two modes of use: one uses the codes; the other is referred to as codeless which means it uses the noise of the system. The disadvantages of codeless are that the ephemerides and clock synchronization data have to be obtained externally and needs both pre- and postsynchronization with an accurate source clock.

P Code

The P code, with its dual frequency, is the basis of PPS (Precise Positioning Service) for high-precision positioning measurements and can give accuracies of 10 to 20 m. Originally it had the drawback of making information available only at discrete intervals rather than in a continuous manner, but this no longer applies. The P code is at present accessible to anyone who has permission to acquire the appropriate receiver, but ultimately this could be restricted, for reasons of national defense, to use by the U.S. military authorities or subject to only limited civilian access.

C/A Code

The C/A code is the basis of SPS (Standard Positioning Service) for initial signal acquisition and course position determinations, which can give an accuracy of 20-30m.

Y Code

When the P code is encrypted to protect it against hostile imitation, it becomes known as the P(Y) code. This technique is called antispoofing (AS), and a special module is required for the decryption of the Y code. This form of tranmission is restricted to military use.

RANGE VALUES

While it would appear that there are three basic variables to be found for each three-dimensional point position (E, N, and H—or longitude, latitude, and height), there is a fourth unknown in the form of the clock bias or offset. Thus, to solve for four variables it is necessary to have at least four range values to satellites.

In fact, each range from a satellite defines a sphere on which the required point lies, and it is the intersection of such spheres from different satellite positions that gives the result (figure 69). We can visualize this if we imagine a bite taken out of an apple. The outline of the bite is the line of intersection, and if the cut is made by a spherical object, then the outline is a circle. Thus if a third such sphere cuts the other two, it will produce a trisection point at a position on the circle. The uncertainties caused by the clock offset can be resolved by introducing the sphere for a fourth range.

This approach is only capable of a few meters' accuracy, so surveying generally requires alternative methods. These can be differential (relative) positioning techniques that use phase measurement, which is similar to the translocation method as used in the transit system.

To determine the range from ground station to satellite, the receiver produces a code similar to that sent from the satellite. Then by comparison of the actual and duplicate codes it is possible to find the transit time from satellite to ground station. In effect, the measured delay between the two signals is the time of travel from satellite to receiver.

REAL-TIME GPS

This term implies that the GPS data is being processed at the same time as it is being collected, albeit a fraction of time after the event. For this to take place, some reliable method of communication is required for transmitting and receiving the data between the control (reference) GPS receiver and the remote (roving) unit. Initially, real time was used for lower-accuracy applications of GPS. The technology has advanced so that it is now used for applications requiring accuracies of a centimeter or better, particularly some civil engineering setting-out applications.

Application of the velocity of light will give an approximation to the range. It is approximate because at that stage it contains the effects of several errors and so is called the pseudo-range. Not until the errors have been eliminated can the ranges be considered true and used for coordinate determination.

For relative positioning there is a requirement for one station to be known (the master) so that the point in question (remote) can be coordinated in relation to it. Relative positioning also requires simultaneous observations at each point from the same satellites. This allows several error sources to be reduced or to cancel out. The same differential measurement technique has been used to improve the results from EDM and allow millimeter accuracy where normally only centimeters are possible.

The fractional part of a wavelength—or phase shift—can be determined by a phase meter, but the integral number of wavelengths, called the 2π

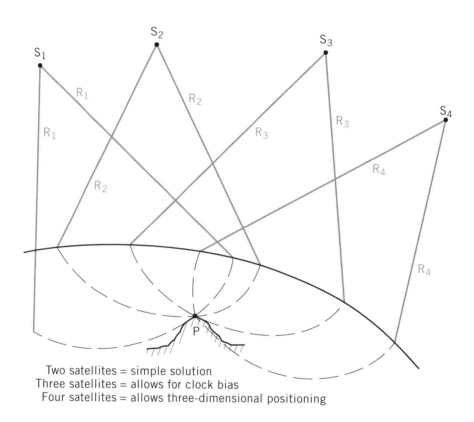

Two satellites = simple solution
Three satellites = allows for clock bias
Four satellites = allows three-dimensional positioning

69. Position circles

ambiguity, can only be resolved by comparisons of signals received at each ground point.

The phase measurement technique is again comparable to that in EDM since the distance is a multiple of the wavelength plus a fraction of a wavelength. EDM, however, records the double distance—to and from the target—which does not apply to GPS.

As in so many other aspects of surveying, extra observations, if in the right configuration, can improve the results. So with the differential technique, increasing the number of pairs of ranges from the same two ground stations can make noticeable improvements.

Over short baselines—those of a few kilometers—errors due to the troposphere and ionosphere can be assumed to be similar at both terminals and will cancel out with appropriate station configuration. For longer separations such errors cannot be considered equal and must be measured and allowed for in some way. With pairs of ranges to two satellite positions, additional error sources cancel out (see figure 64). Such a quadrilateral configuration could be repeated several times during the pass of a single satellite or repeated with other satellites, to strengthen the calculated relative position. Accuracies in the 10-mm region are feasible with this technique.

Data from the satellites gives geographic positions in terms of a particular figure of the earth (WGS 72) or rectangular coordinates on the GPS datum. Hence there may be a need to transform the results into a local system, and for this some knowledge of that system is required. The heights have no direct relationship to mean sea level but can be transformed to ellipsoidal heights and then to orthometric values.

SOURCES OF ERROR

As in all aspects of surveying, GPS is subject to a host of possible errors. While this text is not the place to go into them in depth, some at least require mention. To begin with there are five elements to the whole system, each of which presents its own problems. These are the satellite, the signal and atmosphere, the receiver, selective availability, and the datum.

Satellite

The orbit in which the satellite moves is subject to a range of influences, the predicted effects of which are broadcast by the monitoring stations. In total, their effect is usually less than 100 m or 5 ppm on a baseline vector. The latest satellites are better known than the earlier ones, so this figure should show a decreasing tendency. Among the factors are the variability of the earth's gravitational attraction, and the effects of both the ocean tides and earth tides. These last two are predicted from global models that are still sketchy in parts.

Signal and Atmosphere

Atmospheric effects are well known in various aspects of survey operation. They are similarly an incompletely understood problem with regard to GPS. While the transmission errors are modeled in the software, there are delays that could amount to many meters, due to both the ionosphere and the troposphere. This applies not only to the horizontal component but also to the vertical component of the results. Even with modeling, the height can be in error by several centimeters due to this cause.

Receiver

The receiver is subject to both clock errors and atmospheric effects. Again, modeling of the local situation happens within the software, but codeless-type receivers present more difficulty than coded versions.

Selective Availability

Selective availability (SA) refers to the ability of the U.S. Department of Defense to deliberately degrade the signal for security purposes. The effect can be up to 300 m. It is achieved by manipulating either the navigational message or the satellite clock frequency. Its name arises because authorized (or selected) operators can get access to the unadulterated signals and hence obtain higher accuracies. For those operating in the differential mode, SA does not affect the result. Without SA a position could be determined to within 15 m after just a few minutes of C/A code observations. When SA is turned on, the accuracy drops to the order of 150 m. To achieve the higher figure requires several hours of observation.

Datum

Particularly in relation to height values, there is the uncertainty of the relationship at any point between the geoid and the ellipsoid. The separation of these two needs to be well known if a reliable result is to be obtained. Unfortunately, as with models of other factors, the world is still insufficiently covered with the necessary gravity data to always produce good results for the separation of the surfaces.

THE FUTURE

As the applications for GPS increase, so the inconveniences of selective availability multiply. It seems highly likely that if some of the major operators, particularly in the area of navigation, are not allowed full, unhindered access, then other independent systems will be developed.

CHAPTER
11

Gravity

The effect of the intensity of gravity impinges on many aspects of geodesy. Its most obvious effect is to make a plumb bob hang in a particular manner—often loosely referred to as vertical, although this needs qualification by saying to what it is perpendicular. As has been seen earlier (figure 46), the plumb line can be deflected from the vertical by large mountain masses, and this effect has to be accounted for.

Measurement of the intensity of gravity has been possible for over 300 years and can be either relative or absolute. As with some other aspects of surveying, the relative approach can produce acceptable results far more quickly than absolute measurements.

It is because the earth is both nonspherical and nonhomogeneous that the acceleration due to gravity varies from point to point. Thus it varies with latitude, elevation, the distribution of visible masses and mass deficiencies (such as ocean areas), the distribution of invisible masses, and density variations through the earth.

The standard unit of measure is the *gal*—named after Galileo Galilei— which is equivalent to 10 mm/sec^{-2}, and the magnitude at the earth's surface is in the vicinity of 980 gal. In terms of the usual unit of measure, the milligal (mgal or 0.01 mm/sec^{-2}), variations range over some 5000 mgal due to latitude and 30 mgal per 100 m (9.4 mgal per 100 ft) for changes in height.

ABSOLUTE MEASUREMENTS

The absolute value is normally determined by timing a freely falling body, but high degrees of accuracy were not possible until the advent of electronic timing devices. Prior to this, the tedious but straightforward approach was to time oscillations of a pendulum. The duration of an oscillation directly relates to the length of the pendulum and the acceleration due to gravity, *g*. A length of 25 cm (10 inches) gives a time of 1 second, and 1 m (40 inches) a time of 2 seconds. The latter relation is normally referred to as a *seconds pendulum*, and the former as a *half-second pendulum* since the time in the basic pendulum formula refers to a swing from one limit to the other and back again.

It was Jean Richer's observation of a seconds pendulum in Cayenne in 1672 that first indicated that the earth might not be spherical. He found that the length required to beat single seconds was 1¼ lignes (2.5 mm) shorter in Cayenne than in Paris—suggesting an oblate earth. To determine the length accurately, oscillations were observed over many hours so that the deduced time for a single swing was of the order of 1 ppm.

This method persisted until this century with refinements in the structure of the pendulum; an example is Henry Kater (1777–1835), who in 1818 developed a reversible pendulum with two knife edges. Absolute values were increasing in importance and various methods were devised to measure falling bodies in such a way as to obtain accurate results. One such invention involved photographing the lines on a free-falling scale. Other systems used lenses and mirrors, a falling glass sphere, falling corner cube reflectors, and symmetrical free motion of an object as it was projected up and allowed to fall again.

RELATIVE MEASUREMENTS

The procedure for relative measurements of gravity is again rather like the single-base barometric heighting method. When pendulums were used, measurements would commence at a point of known absolute gravity value. Then the pendulum would be taken to each required position in turn and the period of swing compared with that at the known station to obtain the gravity value at the new point. Today the gravimeter has replaced the pendulum. A *gravimeter* (or *gravity meter*) is any instrument in which the acceleration due to gravity is determined by measuring the force necessary to support a mass in the earth's gravity field. The principle of a gravity meter is based on the fact that when a weight hangs from a spring, the spring changes length with changes in gravity. For example, a nearly horizontal beam can be maintained in position by a spring under tension. As gravity changes, the beam tends to move, so the restraint exerted by the spring can be recorded. In operation a dial reading representing the force necessary to support the test mass is made first at a base station and again at the desired location. The difference in dial readings between the two stations represents the change in gravity. Changes as small as 0.01 mgal can be readily detected, and even 0.001 mgal is possible.

Gravity meters can be used at sea or in aircraft as well as on land. Using them in vehicles, however, creates added problems. It is necessary to eliminate the effect of the moving vehicle since the value required is a stationary one for a particular coordinate position. The effect of motion is called the *Eötvös effect* after the Hungarian scientist Baron von Eötvös (1848–1919), who from about 1886 wrote extensively on gravity. The speed of the vehicle, its direction of travel, and the latitude are all needed to evaluate the correction.

MEASUREMENT USE IN GEODESY, GEOLOGY, AND GEOPHYSICS

Since observed variations in the gravity field are caused by variations in the density of the underlying rock, geologists and geophysicists frequently use gravity information in the analysis of crustal structure. Just the opposite process is used by the geodesist, who uses the crustal structure to indicate the gravity for the area. For example, correlations between known structural and gravity variations can be used to estimate gravity anomalies for areas where no observed values of gravity are available but the crustal conditions are known to be similar. Correlations can also be used to evaluate the reliability of observed gravity anomalies.

The use of geological information to estimate anomalous gravity is probably just a prelude to a much wider use of both geological and geophysical data for geodetic purposes. Discoveries in the last 30 years indicate possible modifications in classical geodetic concepts regarding the relationships between topography and crustal structure. The use of such data may also result in more precise methods of reducing observed gravity to sea level.

CHAPTER
12

The Gyroscope

As a survey tool the gyroscope is probably best known as a unit that can be attached to a theodolite in such a way as to allow determination of the meridian direction. In this form it is, for example, ideal for carrying orientation through a tunnel, although it can be just as usefully applied to orientation problems on surface surveys.

For some years the same type of unit has also played a prominent role in, and developed from, navigation for both missiles and aircraft, so brief details of its operation will be relevant prior to describing its use in inertial surveying.

The essential parts of a gyroscope are as follows:

1. A heavy balance wheel in which most of the mass is concentrated toward the rim. It is made to spin at very high speeds.

2. An axle to support the balance wheel, where the axle's bearings are as nearly frictionless as possible.

3. A gimbal system in which the axle is mounted so as to allow the wheel to rotate freely and take up the appropriate orientation.

As the wheel spins, it is constrained so that its spin axis remains horizontal and rotation occurs only around the vertical axis. The torque exerted upon it makes the spin axis seek true north, and it will oscillate about this direction in a simple harmonic mode.

When a gyroscope is attached to a theodolite (in which form it is known as a *gyrotheodolite*), it is possible to record the angular position of the system in relation to a reference object some distance away. That is, the direction from the instrument to the reference object can be related to true north.

Although the axes in a theodolite system are restrained, even when the gyroscope is free, it will maintain its plane of rotation irrespective of how the supports move. With good conditions and sufficient time it is possible to record with accuracies of a few seconds of arc.

While the use of a gyrotheodolite is somewhat limited it is the application of a gyroscope in an inertial surveying system (Chapter 13) that is of growing importance.

CHAPTER
13

Inertial Surveying

The use of the gyroscope as an aid to navigation can be traced to the early years of the century, but the most notable military application was for missile guidance in the V2 rockets of World War II. As a tool for the surveyor it is of much more recent date. Two firms were given U.S. government contracts in the early 1960s for military positioning systems, and by 1967 one of these firms was asked to develop the equipment for survey purposes. Not until the mid-1970s was a high degree of accuracy achievable, as a development from guidance equipment needed for the space race.

Inertial surveying is a completely new method of positioning for the surveyor. In essence it consists of a unit (albeit a rather bulky one at present) that is transported by truck or helicopter between a series of points whose positions are required, rather along the lines of a conventional theodolite traverse, although successive points do not need to be intervisible. Thus, by starting at a known position and closing at the same, or other known position, the intermediate stopping points can be coordinated. Repetitions of a circuit can improve the results appreciably. The coordination is found in terms of the amount by which the unit has moved from point to point in three dimensions. To achieve this requires three accelerometers (see below) and three gyroscopes mounted in gimbals on a stable base. They must be rigid in space—that is, maintain their orientations against attempts to move the spin axes from the original positions.

ACCELEROMETERS

An *accelerometer* is basically a sprung mass arranged in such a way that it moves as the vehicle moves and has to be counteracted by a restoring force to return it to its original position. The amount of restoring force is a measure of the acceleration, which can be converted into the equivalent distance traveled.

An accelerometer is capable of recording both acceleration and deceleration against time.

A single unit is capable of detecting angular movement in relation to true north. In addition, the equipment can give information on the deviation of the vertical and acceleration due to gravity.

In essence one can use the basic formulae for motion in a straight line to illustrate the operation of inertial equipment. A single accelerometer can detect the acceleration imposed upon it at discrete intervals of time (fractions of a second). Starting from a stationary position, the velocity at the end of an interval can be calculated as the second integral of the acceleration, and from this the distance traveled during the interval is determined. In its turn this positions the instrument with relation to its starting point.

If instead of one accelerometer there are three mounted orthogonally, then the three-dimensional (rather than simply the one-dimensional straight line) movement can be determined. This description is somewhat of an over-simplification since, as with most survey operations, there are side effects to consider. Examples of these are the Coriolis effect, variations in the gravitational field, and deviations of the vertical.

Several methods are available to keep the axes of the accelerometers maintained in a coordinate framework as defined in terms of a geodetic reference system.

In addition to a vehicle moving with respect to the reference system, the earth itself is also in motion. This effect is overcome by the continuous application of torque on the gyroscopes. When an outside force tries to move the axis from its alignment, the effect will be to make the wheel precess in a direction at 90° to that of the applied force. Since it maintains its orientation, it is said to be *rigid in space*. This retention of orientation in space continues as the earth rotates, such that the axis of the gyroscope tilts by measurable amounts through a full circle in 24 hours. If this idea is extended to three gyroscopes mounted at right angles to one another, then it follows that movements can be detected in all three direction.

The combination of three gyroscopes and three accelerometers allows measurement of the directions and magnitudes of movement as well as stabilization of the whole system.

ZERO VELOCITY UPDATE

One particular problem is that an accelerometer cannot separate accelerations due to movement from the acceleration due to the earth's gravitational field. This can only be overcome by periodic stops (every 3 to 5 minutes) for about 20 seconds to achieve *zero velocity update* (ZUPT). This eliminates accumulated errors and allows the velocities of each accelerometer to be recorded. The Z accelerometer value will then be almost all due to gravitational effects.

With a land-borne system, regular ZUPT stops should present no problem, but with a helicopter system it may be necessary to use a hover period if landing is not feasible.

Prior to being moved from the starting station the equipment is set running for up to an hour. This allows time to calibrate the system; it can sense the vertical direction at that station, orient to astronomical north, and perform other calibrating exercises. Because the instrument is transported, there is an update of acceleration data every 0.016 sec that, when doubly integrated into E, N, and H changes, enable the axes to be reoriented. When movement ceases, to allow for ZUPT the vertical direction at that point can be located as before, and from this can be found a measure of the deflection of the vertical. At the same time various calibration parameters can be updated.

The data acquired and computations carried out during a traverse are recorded on tape to allow subsequent additional adjustments and conversion to any required coordinate system.

The precision of a gyroscopic system is, unlike that in a traditional traverse, not so much a function of the distance covered but more an inverse function of the time taken. Errors can arise from the mechanics of the system, external effects such as gravity, and from the operator.

CHAPTER
14

Velocity of Light, EDM, and Laser Ranging

VELOCITY OF LIGHT

The basis of much of the application of electronics to surveying is a good knowledge of the velocity of light or radio waves. When taken in a vacuum the two are the same, but when used in the free atmosphere, they are affected slightly differently. In particular, humidity has a greater affect on radio waves than on light waves.

Basic Relationships

If such waves are used to measure distances on the earth or ranges of such bodies as artificial satellites, the basic relationship is that

$$distance = time \times velocity$$

Before such an approach could be used at levels of accuracy required for surveying, the velocity had to be known to better than ± 1 km/sec. Until the 20th century such accuracies were impossible to achieve, so before any serious attempt could be made to use the relation as it stands, it was first investigated the other way around, as

$$velocity = distance/time$$

and extensive experiments were carried out to obtain a reliable value for the velocity of light waves in a vacuum.

Early History

The fact that light waves travel at a finite velocity was appreciated some three centuries ago. During 1676 Olaf Roemer, the Danish Astronomer Royal, was observing the eclipses of Jupiter's satellites and determined that the time taken for light to travel a distance equal to the diameter of the earth's orbit equated to a velocity of light equivalent to 214 000 km/sec. Fifty years later James Bradley, Third Astronomer Royal at Greenwich, deduced a value equivalent to 301 000 km/sec.

The 19th century saw determined experiments to improve the value by optical-mechanical methods. D. F. Arago, director of the Paris Observatory, experimented in 1820 with a rotating mirror. This technique was modified in 1849 by H. L. Fizeau. He transmitted a pulse of light to a distant mirror, and on its return it was interrupted by a rotating cog wheel. At a particular velocity of the wheel the returning ray would be intercepted by the cogs and not be visible to an observer near the source. Using a cog wheel of 720 teeth the first eclipse was found at an angular velocity of 12.6 rps (revolutions per second)—equivalent to 313 300 km/sec for the velocity of light.

In 1862 Jean-Bernard Foucault used a mirror rotating at 500 rps on a 20-m baseline to obtain a velocity of 298 000 km/sec. He was followed by Albert Michelson, who made many experiments over 40 years—from 1879, when he obtained the value 299 910 $\pm$ 50 km/sec; to 1926, when he experimented over a 35-km baseline in the United States to get a value of 299 798 $\pm$ 4 km/sec.

Electrooptical methods first developed around 1925 when the cog wheel was replaced by a Kerr cell between two Nicol prisms. The result was a value for the speed of light of 299 778 $\pm$ 20 km/sec.

The year 1941 was a turning point when Erik Bergstrand of the Geographical Survey of Sweden conceived a blinking light system. Here the cog wheel was replaced by light pulses of known frequency and variable intensity projected over the line and returned to a receiver near the transmitter. In 1947 his tests over 7734 m gave a velocity of 299 793.9 $\pm$ 2.7 km/sec.

At about the same time L. Essen and C. Aslakson were achieving comparable results, but the firm of AGA was so interested in Bergstrand that they aided his research. In 1948 the first tests, made with a prototype Geodimeter, derived a velocity value of 299 793.1 $\pm$ 0.26 km/sec. The results were now so encouraging that the whole concept could be turned around: the Geodimeter was then used to measure distance with the known velocity of light.

Subsequent years have seen various refinements, all of which give results in the region of 299 792.458 $\pm$ 0.001 km/sec. The surveyor was now confidently able to use velocity and time to determine distance. (The accuracy with which time could be recorded had long been in advance of requirements for routine survey operations).

ELECTROMAGNETIC DISTANCE
MEASUREMENT (EDM)

From the late 1940s it has been possible to measure distances from a few tens of meters to over 100 km, quickly and accurately at the press of a few buttons and twiddles of a few knobs on a black box.

The principle of EDM, as stated above, is that distance = velocity × time.

Thus, from the known velocity of electromagnetic waves in a vacuum (about 300 000 km/sec), suitable corrections for the fact that it is not actually in a vacuum, and a measured value for the time to send a signal from one end of a line to the other and back again, it is possible to determine the distance. In other words, the measurement is actually one of time rather than distance.

Almost all EDM instruments operate on phase difference techniques. Any signal radiated to a far point and back again will exhibit a difference in phase angle on its return. This difference can be measured, but to go with that is a need for knowledge of the integral number of wavelengths traveled.

For microwave systems the number of wavelengths is usually found by increasing the measuring wavelengths by factors of 10. An alternative is to use several specific wavelengths. For light-wave instruments two frequencies suffice because of the shorter maximum range to be resolved.

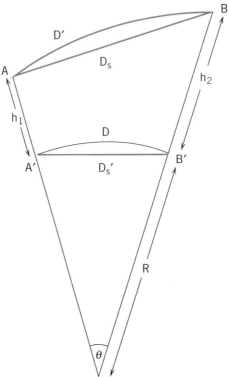

D' = recorded value
D_s = corrected slope value
D_s' = reduced to sea level chord
D = ground equivalent at sea level

70. Reduction of EDM Measurements

All measured distances are inclined lines between elevated terminals and require correction for this as well as for their height above sea level or equivalent reference surface (figure 70). In modern instruments the inclination can now be allowed for automatically.

Instruments that use visible light operate on wavelengths as short as 0.56×10^{-6} m, whereas those that use microwaves are on wavelengths ranging from a few millimeters to 100 or 200 mm.

For light-wave instruments the light source can vary from an ordinary 6-volt bulb to a laser; as a result, the range of operation can vary from a few kilometers to some 50 km.

For microwave instruments the signal is generated by an oscillator which can be much more directional than a light source and has the added advantage of operation in poor weather conditions.

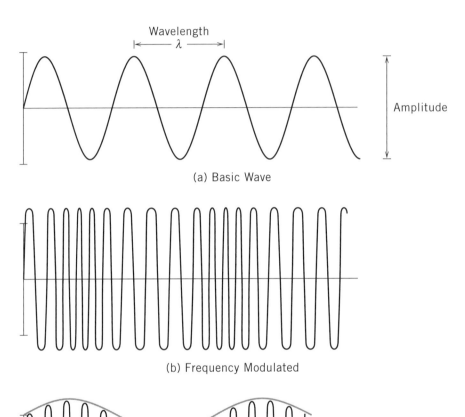

(a) Basic Wave

(b) Frequency Modulated

(c) Amplitude Modulated

71. EDM Waves

Microwave systems use a frequency modulation (figure 71b), particularly as an aid to resolving ambiguities in the recorded distance. On the other hand, the light-wave systems normally use amplitude modulation (figure 71c).

Both forms of instrument are particularly affected by changes in pressure, temperature, and humidity, but corrections are possible as long as meteorological observations are recorded at the same time as the distance. Accuracies of 1 in 100 000 and better are possible depending on conditions.

Instruments of much longer wavelength are used for marine navigation. They are of lower positional accuracy but sufficient for the purposes they are put to. If better accuracy is required offshore, nowadays a satellite fix would be used instead.

The arrangement for long-wavelength positioning is akin to the resection problem. Three fixed land stations are used in adjacent pairs to generate two families of hyperbolae. The signals received indicate differences in distance from the master station and each slave station. Then from knowledge of the number and separation of successive hyperbolae it is possible to determine a position for the receiver.

LASER RANGING

As long ago as the 1950s it was thought feasible to use lasers to determine the distance from the earth to the moon. To put this idea into practice required a reflector system on the moon and this was achieved initially during the Apollo program.

Ranges to the moon were then taken from various purpose-built stations around the globe. The separation of these had little effect on the accuracy of the range measurements which were considered to be in the centimeter region. However, a reflector at such a range is an extremely small target and the procedure was difficult.

Modifications of the same idea are used now on the orbiting satellites. Several passive satellites such as LAGEOS (launched into a 6000-km-high orbit in 1976) and STARLETTE were specifically equipped with reflectors to act as laser targets. They make an easier target than a moon reflector. Even a single-range measure can be in the order of a few centimeters for accuracy, although generally it still requires several weeks of observations to get to this figure.

When a pulse of intense light is aimed at the satellite, some of its photons return. As with EDM, use of the velocity of light together with the transit time gives the distance. If several ground stations are used, it is possible to determine both the orbital parameters of the satellite and the separation of the stations.

Such systems are proving to be of particular use to the geologist and geophysicist for monitoring crustal movements, continental drift, and changes in the earth's parameters.

As with all other systems, a relative mode can be more accurate than an absolute mode, and this is of particular significance in geotectonics. Accuracies in the order of 1×10^{-7} of the separation are possible.

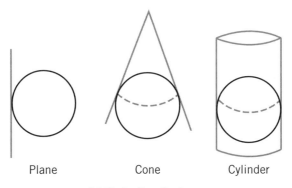

Plane   Cone   Cylinder

(a) Projection Surfaces

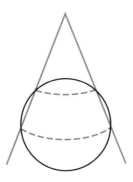

(b) Conic with Two Lines of contact

72. Map Projections

CHAPTER
15

Projections

Once points have been coordinated on an ellipsoid of known size and in relation to an acceptable datum, there remains the problem of depicting those positions in a readily interpreted graphical form, which can be much more meaningful than sheets of numbers.

All through this text, reference has been made to a nearly spherical surface on which all points are positioned. The problem then is to graphically represent a spherical surface on a flat sheet of paper—an exercise referred to as *map projections*. Despite centuries of effort (Ptolemy, A.D. 100–178, produced one of the earliest) no method has yet been found to create map projections without some forms of distortion inherent in the results. These will vary in form and magnitude according to the size of the area involved, the scale at which it is to be represented, and the projection method adopted. It is a vast subject that can only be touched upon here.

Although there are only three basic projection surfaces, by mathematical manipulations it is possible to obtain some 200 variations. Consider an orange where the peel represents the topography that is to become the map. If a small piece of peel is taken off, say a cm square, and laid on the table, it can be flattened out with very little distortion. If, however, a piece several centimeters across is taken off, when pressed flat it will split as well as not lay properly flat. This is an example of the distortion problems found in map projections.

The distortions are of three basic forms—in area, shape, and scale (or distance)—and the respective projection groups are *equal area* (*equivalent*), *conformal* (*orthomorphic*), and *equidistant*. The group most useful for the surveyor is the conformal, in which the scale of small areas is sensibly the same in all directions.

The three basic projection surfaces are the plane, the cone, and the cylinder. The latter two are said to be developable since they can be cut and unrolled to lay flat. Figure 72a shows each in contact with a globe. Consider the simplest

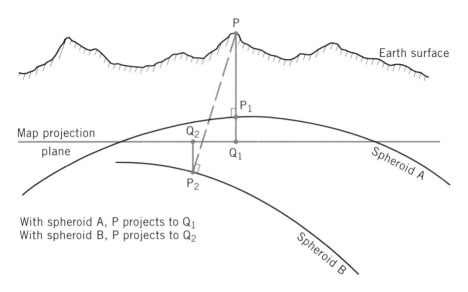

Map projection
plane

With spheroid A, P projects to Q_1
With spheroid B, P projects to Q_2

After Calvert 1994

73. The Effect of Using Different Spheroids

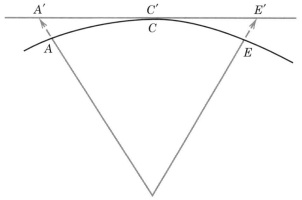

(a) Single point of contact

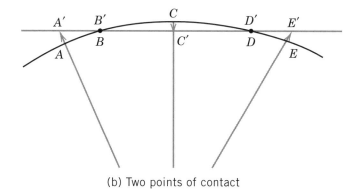

(b) Two points of contact

74. Transverse Mercator Projection

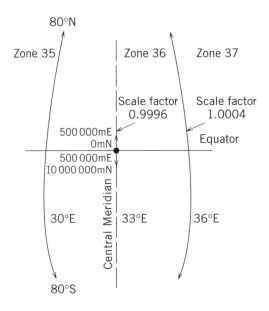

75. UTM Zone 36

case of the plane. Since it touches the globe at a single point, everything on the surface around that position could be transferred to the plane with negligible distortion. The farther one progresses from the point, the greater the distortion. The use of different spheroids can also be a problem (figure 73).

Imagine that the globe is transparent with a point source of light at the center (figure 74). Then points such as *A, B, C, D,* and *E* will be projected by the light source onto the paper plane at *A', B', C', D',* and *E'.* This is the essence of all map projections.

To get different projections, the position of the supposed light source can be varied; the position where the plane, cone, or cylinder touches the globe can be varied; and these surfaces can be made to cut, rather than touch the globe. Add in projections that are mathematical contortions and entail choices as to which distortions to limit and which to accept and one has an almost infinite selection of projections.

WORLD PROJECTION

We have only to look at an atlas to see that there are many ways of representing the whole world on one projection. However, these mostly contain quite large distortions of one form or another.

As the need for a world projection for geodetic work became significant during the 1940s, a specification was drawn up by the U.S. Army to cater to military global requirements. The need was not for an atlas map, which would be of little advantage to soldiers, but for a system on which the whole world could be represented on sheets at a large scale.

There was a need to minimize erros in azimuth, to keep scale errors within prescribed limits, to use as few slices or zones as feasible, and with straightforward conversions of coordinate values from one part of an overlap to the other.

From such requirements grew the *Universal Transverse Mercator (UTM) projection* (sometimes called the Gauss-Krüger projection), which, as the name implies, is a modification of the Transverse Mercator.

It was found that the optimal width of a zone was 6° of longitude—we can liken this situation to an orange with 60 segments. The UTM is a cylindrical projection in which the cylinder can be imagined to have a radius less than that of the globe. Hence there are two lines around the globe that coincide with the cylinder. This particularly keeps the scale error across a zone within the acceptable limits of $\pm 1/2500$. Thus sections *AB* and *DE* (figure 74b) have to be stretched by up to $+1/2500$ to *A'B'* and *D'E'* on the plane, while *BCD* has to be squashed by up to $-1/2500$ to *B'C'D'*.

Every zone (figure 75) is treated identically so that there is a consistent referencing system, and metric units are throughout. In addition, each zone has the same coordinate values on the central meridian—an Easting value of 500 000 m to ensure that all coordinates remained positive, a Northing value of 0 m for working north of the equator, and 10 000 000 m for zones south of the equator.

The zones extend north-south from 80° N to 80° S. Zone 1 is centered at 177° W, so it extends from 180° W to 174° W, and the zone numbers increase to the east. Various ellipsoids were recommended for different parts of the world: Clarke 1880 for Africa; Clarke 1886 for North America; Bessel for the USSR, Japan, and parts of Southeast Asia; Everest for India and neighboring parts of Southeast Asia; and the International 1924 for the remainder. These will be seen to agree closely with the preferred datums (page 95).

MERCATOR PROJECTION

The *Mercator projection* is based on a cylinder with its axis parallel to that of the earth's rotation. Variations on this can be obtained by having the two axes either oblique to one another or at right angles to one another (see "Transverse Mercator Projection" below). If they are oblique to one another, the line of contact is neither a parallel nor a meridian.

The scale varies with distance from the parallel of contact between the cylinder and the earth sphere—which is the equator. Hence it is best for areas near the equator since distortions increase rapidly the farther one gets north or south. No other line of contact can be made between a sphere and cylinder with parallel axes.

Where a country such as Malaya is greater in extent in the north-south direction than east-west but its orientation departs noticeably from the meridian, then an oblique (or rotated) Transverse Mercator is possible.

TRANSVERSE MERCATOR PROJECTION

In the *Transverse Mercator projection* the scale varies with distance from the central meridian; hence it is suitable for areas of limited east-west extent or for areas that can be readily subdivided into separate systems (or zones) in that direction.

This is based on the cylinder with its axis at right angles to the earth's axis of rotation. However, the cylinder can be rotated such that the two axes are still retained at right angles to one another but the line of contact is a different meridian line. With just one line of contact, the scale error will often increase too rapidly for the required area to be covered satisfactorily. By modifying the system to have two lines of contact—achieved by imagining the enclosing cylinder to have a smaller radius than the enclosed sphere—the increase in scale error can be kept within acceptable limits over a far wider extent (see figure 74).

This is the system used in Great Britain. The scale error varies between +1 in 2500 and −1 in 2500, with two meridians where the scale is exact. In Georgia (U.S.A.), where there is a similar system, the range of errors is between +1 in 10 000 and −1 in 10 000.

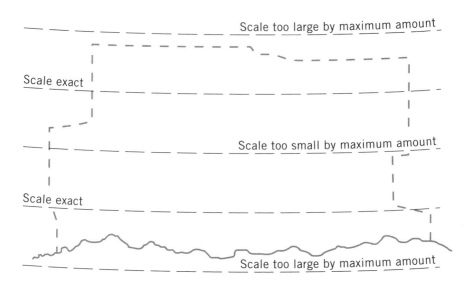

Scale too large by maximum amount

Scale exact

Scale too small by maximum amount

Scale exact

Scale too large by maximum amount

76. Lambert Grid—Typical State or Country Greater East-West than North-South

LAMBERT CONICAL ORTHOMORPHIC PROJECTION

In this projection the scale varies with distance from the central parallel. Hence it is suitable for areas of limited north-south extent or for areas that can be readily subdivided into separate systems in that direction.

As the name implies, the *Lambert Conical Orthomorphic projection* is based on a cone fitting over the sphere or cutting into it (figure 72b). In the former there is contact along one parallel, in the latter along two. Along these standard parallels the scale is correct, and as one moves north or south from these parallels, the error varies. In the case of two standard parallels, the scale is too small between the parallels, and too large outside of them (figure 76).

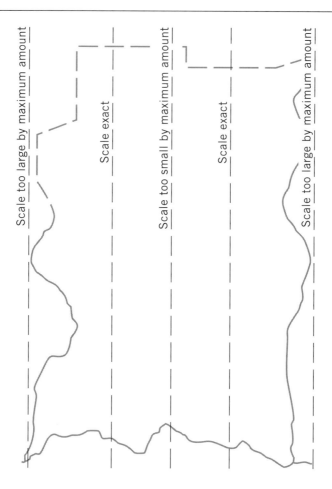

Scale too large by maximum amount

Scale exact

Scale too small by maximum amount

Scale exact

Scale too large by maximum amount

77. Transverse Mercator Grid—Typical State or Country
Greater North-South than East-West

For example, in Connecticut the scale varies from being too small by 1 in 59 000 to being too large by 1 in 34 000, whereas in North Carolina the range is from 1 in 7900 too small to 1 in 5950 too large (see figure 77).

STATE PLANE COORDINATES

In a country as vast in extent as the United States, each state has its own system. Because of the variety of shapes of the states, some 30 have adopted the Lambert system and 19 the Transverse Mercator. It is fortuitous that virtually all the states are oriented more or less either east-west or north-south, and as a result only two different projections are required.

By using a conformal projection as the basis and limiting one direction of a single zone, these maps are able to preserve correctly angles defined by lines of less than about 10 miles (16 km). Moreover, the scale factor, or deviation of the grid length (the distance on a plane) from the geodetic length (the surface distance on the spheroid), can be limited to within acceptable bounds.

One particular problem raised in addition to the scale factor is known as the *convergence of the meridians:* except at a few specific positions, there is a difference between grid azimuth of a line and its geodetic value, and this difference increases the farther the line in question is from the central meridian of the system. The problem arises because, as is well known, the geodetic meridians converge toward the poles whereas the grid equivalents do not.

In the 1930s an engineer submitted to the USCGS a proposal for an arrangement whereby geodetic data could be used over an entire state by applying only the formulae for plane surveying. This resulted in 1933, in the establishment of a coordinate system in North Carolina in which latitudes and longitudes of points throughout the state could be transformed into plane rectangular coordinates on a single grid. Then any surveys within the states could be defined in terms of their coordinates on a common origin.

CHAPTER
16

Examples of
Modern Projects

SUPERCONDUCTING SUPER COLLIDER (USA)

A recent example of the requirement for high-accuracy geodetic control was the Superconducting Super Collider (SSC) in Texas. Although the project terminated because of financial constraints, enough of the work was completed to illustrate the care and accuracy required for such a structure. The main ring of the structure was to be a tunnel 4 m diameter and 87 km long. This tunnel would contain 12 000 magnets, each of which had to be aligned to better than 1 ppm in a perfect geometric plane, not a level surface. The survey tolerance for the excavation, both horizontal and vertical, was 76 mm. The added stricture here was that excavation had to meet the specification precisely at all times and with all parts; no realignment for corrections would be possible.

The design coordinates of the magnet positions were given in an arbitrary site reference system as cartesian coordinates. To transform these into geodetic coordinates required an accurate model of the relative geoidal undulations in the area. Transformed ellipsoidal heights had to be corrected by the geoidal separation to obtain the orthometric heights. It was expected that the relative geoidal heights could be determined with an average standard deviation of 5 mm and the components of the relative deflections of the vertical to 0.1″ of arc.

In addition, a special conformal map projection was defined for the area of the project.

Surface control was designed for high-precision GPS aimed at achieving 3 mm + 1 ppm for the baseline components. The relative height difference across the ring was to be 10 mm. To achieve this, scientists determined the effect of tidal accelerations on the earth's equipotential surface, but found it to be less than 1 mm. Additionally, account was required of the nonparallelism of orthometric surfaces to the reference surface, and this was modeled.

The possible effect of lateral refraction in the tunneling survey was of concern since even a small temperature gradient of 0.2° C/m over 500 m would deviate the line by 23 mm. It was possible to reduce this effect by using gyrotheodolites instead of normal traverse angle measure.

The actual achieved accuracy of the GPS control was ±3mm $\pm\ 10^{-7}$ S. The tunnel was to be connected to the surface by a number of vertical shafts spaced about 4 km apart.

At the time the project closed, 24 km of tunneling was completed. The breakthrough of individual 4 km segments between the shafts was accurate to within a few millimeters.

For further details on the SSC, see Chrzanowski et al. (1995).

GOTTHARD TUNNEL (SWISS ALPS)

Long, deep tunnels present many problems that relate to geodesy. One of the latest such tunnels is the proposed Gotthard base tunnel in the Alps. At 56.9 km long, it will pass between Erstfeld and Biasca. Although a similar length to the Channel Tunnel between England and France, the problems in construction are quite different. While the Channel Tunnel is only 100 m below sea level, the Gotthard tunnel will have up to 2300 m of overburden, which is expected to create tunnel temperatures of 55°C.

From the geodetic point of view, one particular problem is the geoidal-ellipsoidal separation. This will be critical to achieving an accurate break-through. The whole proposed route of the tunnel has been given a geodetic control network using both traditional (triangulation and traverse) and modern (GPS) methods.

A good mathematical model based on the ellipsoid and geoid, with deviations of the vertical and geoidal undulations, was essential for transforming coordinates from the GPS geocentric version to the grid systems; and for correcting both the gyro-azimuths and the astronomical observations. While the geoidal model is well known on the ground surface, geodesists can only assume that the model for the depths under consideration takes a similar form—an assumption that could introduce errors. In addition, the change in gravitational acceleration affects the height difference that is used for the depth between the surface and a shaft bottom some 800 m deep.

During survey of the tunnel operations, refraction will be a particular problem because of the expected large temperature gradients across the tunnel. In the less rigorous conditions of the Channel Tunnel this was found to have a noticeable affect, so it is likely to be a much greater distraction in the Gotthard tunnel.

As far as angular measures are concerned, normal traverse-type angles relate to the geoid while angles from the gyrotheodolite refer to the ellipsoid.

The difference between them is the deviation of the vertical; if this correction is neglected in this situation a large systematic error could result.

Despite all the likely difficulties the expected breakthrough on the most difficult section is 10 cm.

For further details on the Gotthard Tunnel, see Egger (1995).

MINERAL EXPLORATION IN KAZAKHSTAN

The mineral wealth of Kazakhstan and neighboring areas of the Caspian Sea is in the process of development and expansion. So vast is the potential that one of the first requirements is up-to-date mapping. The local geodetic system used a Pulkovo 1942 datum. One of the prime mapping needs was to transform coordinates from the WGS 84, which was done using a network of seven geodetic monuments around the north end of the Caspian Sea. These were located in absolute position to better than 0.5 m. An intricate manipulation of various datum shifts was then required to relate Pulkovo 1942 to WGS 84. It was then possible to calculate the geoidal undulations for the area. While these agreed well in one part of the country, there was an unexplained degradation in other areas. Further gravitational data and a new geoidal model could well iron out the differences. High accuracy is required in this project because of the requirements of the seismic operations.

For further details on mineral exploration in Kazakhstan, see Nash et al. (1995).

HEIGHTING BY GPS IN LESOTHO

In traditional survey work, heighting over long distances has always been a problem because of the refraction effect on the grazing rays as they pass through the lower layers of the atmosphere. Even over the steepest of sights by trigonometric heighting, the vertical angle is seldom more than a few degrees.

GPS, on the other hand, concentrates on elevations as near to the vertical as possible and so greatly reduces the refraction problem. In fact, only data from satellites above a given angle of elevation (such as 10°) are accepted in the solution. For heighting only, it is feasible to concentrate on those satellites orbiting closest to the vertical. Since usually it is differences in height that are required, the geodesist can assume that nearly vertical observations through the atmosphere from points not too far apart contain comparable refractive errors that cancel out—a situation that cannot be so readily accepted in traditional trigonometric heighting.

A fine example of the advantages of GPS is instanced in a paper by K. N. Greggor relating to the Lesotho Highlands Water Scheme. It graphically illustrates a number of the topics discussed elsewhere in this text and demonstrates the need for their recognition and application.

In this scheme there are to be five dams, two power stations, and over 200 km of tunnel. The required control points are located in deep valleys since the tunnels on the project are some 1000 m below the highest summits of the Highlands. Using traditional methods to fix such points numerous mountaintops would be occupied for the control before coordinating the valley points—a process that would have taken several months. By using GPS this has all changed and the actual valley points were occupied and the 70 km of check survey completed in only two weeks.

Height differences measured from satellites differ noticeably from those sensed by flowing water or spirit leveling. In fact, over the 70 km length of the Lesotho section of tunnel, the height differences measured by spirit leveling and those by GPS differed by as much as 2.5 m. The cause could be attributed to the variations in gravity resulting from the variations in internal density of the earth.

A further complication arose in that the national height network refers to the Clarke 1880 ellipsoid, fitted in such a position and orientation as to form the Cape datum. However, this does not coincide with the GPS ellipsoid of WGS 84. Besides differing in dimensions they depart from one another by some 300 m. In addition, there appears to be a small tilt of the order of $1''$ in alignment between the Cape and WGS 84 datums and there may also be a very small scale difference.

In traditional precise leveling it is neccessary to account for the corrections between orthometric and measured heights which are normally very small indeed. In extreme conditions, as in Lesotho, this can amount to several centimeters over 100 km, and with tight fluid-flow parameters this could be a critical amount.

On the other hand, although the effect of the geoidal undulation is reckoned in meters, it is usually adequate to assume that in a local region both the ellipsoid and the geoid are more or less equivalently ellipsoidal in shape although slightly tilted with respect to one another. This tilt is the deviation of the vertical, and in extreme cases can amount to $30''$.

Some might ask why we should not just use GPS heighting and ignore the geoid. Unfortunately, particularly in projects such as the Highlands Water Scheme in Lesotho, there are difficulties with this approach, not least of which is that it is not possible to use satellite observations in tunnels. Thus the GPS height values can only be used on the surface to check the precise spirit leveling as it emerges from underground.

A further problem is that fluid-flow calculations would be off by several seconds of arc, even up to $5''$ in Lesotho. The slope of the tunnel has to be kept within close limits such that the water neither flows too fast nor too slow, so a few seconds can be of considerable importance. In a trial on the edge of an anomalous gravitational section of the scheme, GPS measures on a dam deformation scheme were compared with traditionally obtained results. Although the GPS values were accurate to better than half a centimeter, the tilt was found from the comparison to be $11''$.

It cannot be stressed too strongly that the casual user of GPS software should *never* simply assume that the global geoidal data provided is of adequate resolution for particular heighting uses. It must also be remembered that even in areas well covered with gravity measurements the latent unknown relationship between GPS and geoidal height differences is of the order of 1 in 100 000, whereas GPS heighting alone tends toward an order of magnitude better than this.

For further details, see Greggor (1993).

PLATE TECTONICS

In several parts of the world geodetic methods are being used to determine plate tectonic movements of a few centimeters per year. Both satellite doppler and GPS have been used in Papua New Guinea, Greece, and New Zealand.

Geodetically Papua New Guinea is an interesting area because of its large gravity and magnetic anomalies as well as the plate movements. McClusky (1994) quotes Bouguer anomalies ranging from −180 to 200 mgal and a minimum free-air anomaly of −300 mgal. Obviously a research gold mine for the geophysicist and volcanologist such areas also provide the surveyor with opportunities to use the latest technology to quantify the rate and direction of movement of the earth's crust.

In order to do this sort of analysis it is first necessary to provide over the area of interest a network of control points positioned with the highest possible accuracy. Ideally the whole network is reobserved on a regular basis, which may be annually or every 2 or 3 years depending on the availability of staff, equipment, and funds. Further observation points can then be related to the basic control point.

In the Papua New Guinea work comparisons are possible between 1981 data and various observations in the 1990s. Movements are in the range from +1.51 to −1.93 m over lines of several hundred kilometers. It is also possible to compare results by Doppler and by GPS.

In general, the techniques employed have proved to be appropriate to the requirements of the geologists and geophysicists and to reliably record changes of a few centimeters per year.

For further details see McClusky et al. (1994) and Stolz et al. (1982).

DETERMINING HEIGHT OF MOUNT EVEREST

A good example of the complications of the geoid is the determination of the height of Mount Everest. This mountain is some 400 miles (645 km) from the nearest sea, so where would sea level be if extended under the mountain?

For a long while the height was quoted as 29 002 ft, and the question was raised as to whether or not the "2" at the end should be omitted. If left it might imply an enhanced accuracy, but if it were dropped it would imply that the value was only known to the nearest 1000 ft.

Is 29 002 ft an exact height? In the 1920s De Graaff-Hunter used an interesting analogy. The height of the Eiffel Tower is stated to be 984 ft. But what about the legs of the tower and foundations—should not they be included? Then beneath the tower is a complex of radio installations. The point is that, not only the top, but the bottom also needs defining before quoting a height. In the case of an ordinary tower it would be possible to decide what the bottom was and get access to it. With the Eiffel Tower, however, we would need to remember that it changes by some 3 inches (8 cm) seasonally with temperature. With mountains the base is certainly not visible, and because of tectonic movements the height is also changing slightly.

Some of the possible causes of errors in any height measure would likely be of opposite sign and so cancel out; aside from these there was estimated to be an uncertainty of about 27 ft (8.2 m). In general, there are three main factors affecting the accuracy of the height determination of a mountain:

geoidal separation

atmospheric refraction

deviation of the vertical

In the case of Mount Everest, however, these are compounded by:

vast lengths of sights

great differences in height from observing stations to peak

the huge mass of the mountain range

the distance of the peak from the sea

the impossibility of occupying both ends of the line

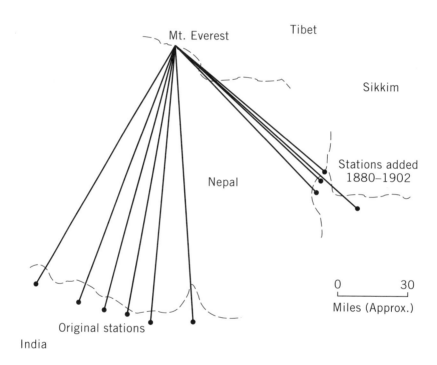

78. Heighting Mount Everest

To summarize the story, Figure 78 shows that because of the political situation in the period 1847–1850 the nearest that surveyors could get to make observations was the Plains of northern India, about 110 miles (160 km) from the peak. Although six observing stations were used, they were far from ideally placed since the maximum intersection angle was 36°. The six results varied by only 36 ft (11m), and the mean value was 29 002 ft. (8840 m). The refraction effects were found to be up to 1375 ft, and an angular variation between morning and afternoon of up to 200″. Such agreement under the circumstances was phenomenal. The biggest sources of error were considered to be:

the effect of refraction with an uncertainty of perhaps 120 ft (36.6 m)

uncertainty in the geoidal separation of perhaps 115 ft (35.1 m)

In September 1992 an international Italian-Chinese expedition remeasured the height. This they did using all the modern techniques of GPS, EDM, and electronic theodolites and related the result to sea level in the Bay of Bengal. Simultaneous observations were made from both Tibet and China with a reflector and satellite receiver on the summit. The value quoted for the orthometric elevation was 8846.10 m. The previous Chinese value of 1975 had been 8848.13 m. The agreement both between the two Chinese values and with the original value of 8840 m is extremely good. It shows that the original work was far better than to the nearest 1000 ft. It also shows that original estimates for the geoidal separation were good.

For further detail see Smith (1997).

CHANGING NATIONAL GRID AND GEODETIC DATUM IN GREAT BRITAIN

Why change to a new datum? There are various reasons, among which Wilson and Christie (1992) give the following:

GPS measurements can now detect any shortcomings of an existing mapping datum.

Existing datums are unique to Great Britain and cannot be extended to cover the rest of Europe.

The need for more of an international datum is overwhelming.

GPS has removed the need for any intervisibility between stations; hence there is now no need for expensive hilltop monuments.

With the rapidly increasing popularity of GPS the datum and related topics have to keep in step.

New applications are arising regularly and require greater accuracies than can be supplied via the existing system.

The first national mapping of Great Britain was based on 13 different meridians on the Cassini projection. Later this was changed to the National Grid based on a modified Transverse Mercator projection. This was later supplemented with a Europe-wide system called the European datum 1950 (ED 50).

The National Grid is based on the Ordnance Survey Great Britain 1936 (OSGB 36) datum, whose origin is at Herstmonceux where the orientation and position of the Airy spheroid is defined. Since 1936 the primary triangulation has been readjusted twice. The first of these resulted in the OSGB 70(SN) datum, in which the coordinates of Herstmonceux were held fixed at their previous values. The outcome of this readjustment indicated that OSGB 36 was not only too large overall, but that it varied considerably in scale from one part of the country to another. The next readjustment—indicated shortcomings in the OSGB 70(SN), and so the process continues.

With the advent of satellite geodesy, GPS positional values were given in terms of WGS 84. However, in Europe there were efforts to improve on this, and a continent-wide campaign of observations took place in 1989. Some 60 receivers were used on 93 stations throughout Europe, and the results were subjected to a rigorous adjustment. The result was the European Terrestrial Reference Frame 1989 (ETRF 89), which is now the fundamental frame throughout Europe. In this, the standard errors attributed to the observing stations are around the 10-cm level and expected to be improved to 1 or 2 cm. What it has shown, however, is that the original OSGB 36 triangulation and its equivalents in other European countries are not compatible with ETRF 89 and that unfortunately the differences are not uniform. As a result coordinates in WGS 84 will differ from coordinates in OSGB 36 by hundreds of meters—due solely to the difference of datums and not to error.

This discrepancy obviously becomes a problem for a national mapping organization. After national consultation a policy was devised to support transformations between the national grid (OSGB 36) and WGS 84 (or ETRF 89). To this end, conversion tables have been compiled and made freely available that will have sufficient (2-m) accuracy for mapping purposes. A more precise transformation (0.2 m) is also available as a service. In tandem with this is a transformation service for situations requiring higher accuracies. For the time being, mapping will be retained on OSGB 36.

A parallel development has been a national geoidal model showing the geoidal-ellipsoidal separations with a claimed accuracy of 5 to 7 cm.

For further details, see Calvert (1994).

Bibliography

For readers interested in pursuing the subject in more depth the following is a selection of further reading. Numerous other papers are available and in particular it is recommended to obtain the proceedings of relevent conferences, seminars and symposia for details of the latest state of this fast changing science.

Abbreviations

ACSM   American Congress of Surveying & Mapping

AFT   Association Française de Topographie

Aust. Sur.   Australian Surveyor

Bull. Geod.   Bulletin Géodésique

Can. Sur.   Canadian Surveyor

C. E. Sur.   Civil Engineering Surveyor

FIG   Fédération Internationale des Géomètres

G.I.M.   Geodetic Info Magazine

N.Z. Sur.   New Zealand Surveyor

P.O.B.   Point of Beginning

RAS Royal Astronomical Society

RICS   Royal Institution of Chartered Surveyors

S. & M.   Surveying and Mapping

Sur. Rev.   Survey Review

S. W.   Surveying World

Aardoom, L. 1972. On a geodetic application of multiple-station very long base interferometry. *Netherlands Geodetic Commission*, new series, 5, No. 2.

Allman, J.S. & Veenstra, C. 1984. *Geodetic model of Australia 1982*. Tech. Report 33. Division of National Mapping, Australia.

Anderle, R.J. 1980. Accuracy of mean earth ellipsoid based on doppler, laser and altimeter observations. *Bull. Geod.* No. 54:521–527.

Anderle, R.J. 1981. Geodetic control from electronic satellite observations. FIG, Montreux. Paper 506.1.

Ashgaee, J. 1986. IC2 technique for GPS signal processing in survey applications. FIG, Toronto. Paper 509.1.

Ashkenazi, V. 1985a. Positioning by satellites: trends and prospects. RAS Mason Conference. Glasgow.

Ashkenazi, V. 1985b. Positioning by GPS and NAVSAT. Will it be the end of geodetic networks? Paper H3. Survey and Mapping '85 (conference). Reading.

Ashkenazi, V. 1986. Satellite geodesy—its impact on engineering survey. FIG, Toronto. Paper 502.1.

Ashkenazi, V. 1994a. The impact of GPS on civil engineering projects: a blessing or a curse? *Proceedings of 3rd CITOP Conference,* London, Dec. 9, 1994. Published by AFT and RICS.

Ashkenazi, V. 1994b. GPS and maps. FIG Congress, Melbourne. Paper TS 505.1.

Ashkenazi, V. & Gough, R.J. 1975. *Determination of position by satellite-doppler techniques.* Univ. of Nottingham.

Ashkenazi, V. & McLintock, D. 1982. Very long baseline interferometry: an introduction and geodetic applications. *Sur. Rev.* 26, No. 204:279–288.

Barr, J.R. & Carriere, R.J. 1981. Accuracy criteria for inertial surveys. FIG, Montreux. Paper 512.2.

Bevin, A.J. & Hall, J. 1994. The review and development of a modern geodetic datum. FIG Congress, Melbourne. Paper TS 506.3.

Billaris, H. et al. 1991. Geodetic determination of tectonic deformation in central Greece from 1900 to 1988 *Nature* 350, No. 6314:124–129.

Bloodgood J.F. 1992. Jumping on the GPS Bandwagon. ACSM Bulletin No. 136 (March/April):32–34.

Bomford, G. 1971. *Geodesy,* 3rd ed. Oxford Univ. Press, Oxford.

Bossler, J.D. & Hanson, R.H. 1984. The impact of VLBI and GPS on geodesy. *S. & M.* 44, No. 2:105–113.

Bossler, J.D. et al. 1980. Using the global positioning system (GPS) for geodetic psoitioning. *Bull. Geod.* No. 54:553–563.

Brown D.C. 1976. Doppler surveying with the JMR-1 receiver. *Bull. Geod.* No. 50:9–25.

Butelli, G. et al. 1994. GPS measurements for ground subsidence monitoring and geoid computation in the Ravenna area. FIG Congress, Melbourne. Paper TS 501.3

Calvert, C. 1994. Great Britain-changing the national grid and geodetic datum. *S. W.,* Sept., pp. 23–24.

Carter, W.E. 1981. An elementary introduction to radio interferometric surveying. *Sur. Rev.* 26, No. 199:17–31.

Cavero, A.P. & Czarnecki, K. 1994. GPS-related reference system as applied to engineering purposes. FIG Congress, Melbourne. Paper SGS 653.3.

Chen, J. Y. 1994. Crustal movements, gravity field and atmospheric refraction in the Mt Everest area. *Zeitschrift für Vermessungswesen* No. 8, pp. 389–400.

Christie, R.R. 1992. The establishment of a new 3D control framework for GB based on modern space techniques. *S. W.,* Nov., pp. 41–43.

Chrzanowski, A. et al. 1995. US Supercollider-geodetic control. *S. W.,* March, pp. 22–26.

Clarke, F.L. 1980. Determination of a local geodetic datum from astrogeodesy. *Sur. Rev.* XXV, No. 196:268–279.

Cleasby, C. 1992. The determination of height using GPS. *C. E. Sur.,* Oct., pp. 25–27.

Collier, P.A. 1994. Static and kinematic GPS for deformation monitoring. FIG Congress, Melbourne. Paper SGS 552.3.

Collier, P.A., Armstrong, A.P. & Leahy, F.J. 1994. GPS heighting by least squares collocation-initial results and experiences. FIG Congress, Melbourne. Paper TA 501.2.

Cook, A.H. 1969. *Gravity and the earth.* Wykeham Publications.

Cross, P.A. 1986. Prospects for satellite and inertial positioning methods in land surveying. Land and Minerals Surveying, April, pp. 196–203.

Cross, P.A. 1991. Position: just what does it mean? *Jnl. Inst. Navigation,* pp. 246–262.

Cross, P.A. & Wood, J.P. 1980. Instrumentation and methods for inertial surveying. *Chartered Land & Minerals Surveyor,* Autumn.

Danson, E.F.S. 1991. Introduction to plane and geodetic coordinate systems. *C. E. Sur.,* July/Aug., pp 28–29; Sept., pp. 11–12.

Deakin, R.E., Collier, P.A. & Leahy, F.J. 1994. Transformation of coordinates using least squares collocation. *Aust. Sur.,* March, pp. 6–20.

Dewhurst, W.T. 1984. Local gravity estimates available from National Geodetic Survey. *ACSM Bull.,* June, pp. 13–16.

Dodson, A.H. 1994. The status of GPS for height determination. FIG Congress, Melbourne, Paper TS 501.1.

Doyle, D. & Dragoo, A. 1994. Where freedom stands. *Professional Surveyor,* Jan./Feb., pp. 4–9.

Egger, K. 1995. The survey of the Gotthard base tunnel. *S. W.,* March, pp. 18–21.

Ezeigbo, C.U. 1994. On the choice of a suitable datum for a unified geodetic network of Africa. *S. African Jl. of Surveying & Mapping,* Dec., pp. 385–392.

Featherstone, W.E. & Olliver, J.G. 1994. A new gravimetric determination of the geoid of the British Isles. *Sur. Rev.* 32, No. 254 (October):464–478.

Fell, P.J. 1980. A comparative analysis of GPS range, doppler and interferometric observations for geodetic positioning. *Bull. Geod.* No. 54:564–574.

Ferguson, J. 1990. How the flat earth got round: an ancient history. *Ontario Land Surveyor,* Summer, pp. 8–9.

Furguson, J. 1991. What is height anyway? *Ontario Land Surveyor,* Spring, pp. 17–19.

Fischer, I. 1974. A continental datum for mapping and engineering in South America. FIG, Washington, September.

Fowler, R. 1992. All about coordinate systems. *Northpoint,* Summer, pp. 20–24; Fall, pp. 24–28.

Frei, E. & Schubernigg, M. 1992. GPS surveying techniques using the fast ambiguity resolution approach (FARA). 34th Australian Surveyors Congress, Cairns.

Frodge, S.L. et al., 1994. Results of real-time testing of GPS carrier phase ambiguity resolution on-the-fly. FIG Congress, Melbourne. Paper TS 606.3.

Garland, G.D. 1977. *The earth's shape and gravity.* Pergamon Press, Toronto.

Gilliland, J.R. 1984. Comparison of height derived from a combined solution of doppler and free air geoid data with AHD values. *Aust. Sur.* 32, No. 3:186–192.

Goad, C.C. 1989. On the move with GPS. *P. O. B.,* Apr./May, pp. 36–44.

Greggor, K.N. 1993. GPS in Lesotho-the heighting problem and rapid height monitoring by GPS. *J. Inst. Mine Surveyors* (S. Africa) 27, No. 3:43–58.

Hajela, D. 1990. Obtaining centimeter-precision heights by GPS observations over small areas. *GPS World,* Jan./Feb., pp. 55–59.

Hannah, J. 1985. The global positioning system—the positioning tool of the future. *N. Z. Sur.,* Aug., pp. 268–281.

Haug, M.D. et al. 1980. A simplified explanation of doppler positioning. *S. & M.* 40, No. 1:29–45.

Helmer, G. 1991. Which way is up in earth orbit? *Professional Surveyor,* July/Aug., pp. 39–44.

Hoar, G.J. 1982. Satellite surveying. Magnavox, Torrance, CA.

Hogarth, B. 1990. The global positioning system. *C. E. Sur. Suppl.,* May, pp. 2–4.

Hollmann, R. & Welsch, W.M. 1994. GPS for precise engineering surveying applications. FIG Congress, Melbourne. Paper TS 606.1.

Homes, G.M. 1992. Changing technology-monitoring sea level change. *Aust. Sur.* 37, No. 1 (March):13–22.

Hothem, L.D. et al. 1984. GPS satellite surveying-practical aspects. *Can. Sur.* 38, No. 3:177–192.

Hui, P.J. 1982. On satellite signal processing techniques applicable to GPS geodetic equipment. *Can. Sur.* 36, No. 1:43–54.

Jackson, J.E. 1980. *Sphere, spheroid and projections for surveyors.* Granada, London.

King-Hele, D.G. 1960. *Satellites and space research.* Routledge & Kegan Paul, London.

King-Hele, D.G. 1963. *The shape of the earth.* Institute of Navigation, London.

King-Hele, D.G. 1969. *The shape of the earth.* Royal Aircraft Establishment, Farnborough.

Kinlyside, D.A. 1994. The geocentric datum of Australia—transform or readjust? FIG Congress, Melbourne. Paper TS 506.1.

Kleusberg, A. & Langley, E.B. 1990. The limitations of GPS. *GPS World,* Mar./Apr., pp. 50–52.

Lachapelle, G. 1986. GPS-current capabilities and prospects for multi-purpose offshore surveying. FIG, Toronto. Paper 502.2.

Langley, R.B. 1990. Why is the GPS signal so complex? *GPS World,* May/June, pp. 56–59.

Langley, R.B. 1992. Basic geodesy for GPS. *GPS World,* Feb., pp. 44–49.

Leach, M.P. & Cardoza, M.A. 1994. GPS monitor stations, reference stations and networks. FIG Congress, Melbourne. Paper TS 504.2.

Leick, A. 1993. Geodesy in review. *ACSM Bulletin,* Jan/Feb., pp. 45–48.

Leick, A. 1994. *GPS Satellite surveying,* 2nd ed. Wiley Interscience, New York.

Levallois, J.J. 1988. *Mesurer la Terre. 300 ans de geodesie française.* Presses de Ponts et Chaussées, Paris.

Lippold, H.R. 1980. Readjustment of the National geodetic vertical datum. *S. & M.* 40, No. 2:155–164.

Macdonald, A.S. & Christie, R.R. 1991. From miles to millimetres: the story of geodesy at Ordnance Survey, 1791–1991. *Sur. Rev.* 31, No. 241:126–147.

Manning, J. & Harvey, W. 1994. Status of the Australian geocentric datum. *Aust. Sur.,* March, pp. 28–33.

McClusky, S. et al. 1994. The Papua New Guinea satellite crustal motion surveys. *Aust. Sur.,* Sept., pp. 194–214.

McLintock, D.N. 1987. GPS-Land Survey applications. RICS/Royal Institution of Navigation meeting. London.

Meade, B.K. 1983. Latitude, longitude and ellipsoid height changes NAD 27 to predicted NAD 83. *S. & M.* 43, No. 1:65–71.

Milbert, D.G. 1992. GPS and geoid 90—the new level rod. *GPS World,* Feb., pp. 38–43.

Moritz, H. 1990. *The figure of the earth. Theoretical geodesy and the earth's interior.* Wichmann, Karlsruhe.

Mueller, I.I. 1981. Inertial survey systems in the geodetic arsenal. *Bull. Geod.* No. 55:272–285.

Mueller, I.I., Hannah, J. & Pavlis, D. 1981. Inertial technology for surveying. FIG, Montreux. Paper 501.4.

Nash, D., Whiffen, P. & Dixon, K. 1995. Surveying Kazakhstan's rich resources with GPS. *GPS World*, Feb., pp. 22–30.

Olliver, J.G. 1981. Satellite-derived geoids for Great Britain and Ireland. *Sur. Rev.* 26, No. 202:161–179.

Olsen, N.T. 1993. Understanding the differences between an ellipsoid, a geoid and a spheroid. *G.I.M.*, May, pp. 64–65.

Ordnance Survey. 1995a. The ellipsoid and the transverse mercator projection. *Geodetic Information.* Paper 1.

Ordnance Survey. 1995b. National grid/ETRF 89 transformation parameters. *Geodetic Information.* Paper 2.

Parkinson, B.W. 1979. The global positioning system (NAVSTAR) *Bull. Geod.* No. 53:89–108.

Pelletier, M. 1990. *La Carte de Cassini.* Presses de Ponts et Chaussées, Paris.

Poretti, G. & Beinat, C.M.A. 1994. GPS surveys on Mount Everest. *GPS World,* Oct., pp. 33–44.

Rapp, R.H. 1994. Separation between reference surfaces of selected datums. *Bull. Geod.* 69:26–31.

Rizos, C., Stolz, A. & Masters, E.G. 1984. Surveying and geodesy in Australia with GPS. *Aust. Sur.* 32, No. 3:202–224.

Schwarz, K.P. 1981. *Error characteristics of inertial survey systems.* FIG, Montreux. Paper 512.1.

Seeber, G. 1993. *Satellite geodesy: foundation, methods and applications.* de Gruyter, Berlin and New York.

Seeger, H. 1994. EUREF: The new European Reference Datum and its relationship to WGS84. FIG Congress, Melbourne. Paper TS 506.4.

Sherwin, T. 1995. A look at some of the main tidal constituents. *Hydrographic J.* No. 75 (Jan.):15–19.

Smith, J.R. 1969. The development of two standards. *Sur. Rev.* No. 153:133–146.

Smith, J.R. 1983. Geodimeter 1947–1983. Geotronics, Huntingdon, UK.

Smith, J.R. 1987. *From plane to spheroid.* Landmark Enterprises, Rancho Cordova, CA.

Smith, J.R. 1996. *Capt. R. S. Webb MBE, RFA (1892–1976): From Shropshire to Paarl via Geodesy and Lesotho.* S. African Council for Professional and Technical Surveyors, Durban

Smith, J.R. 1997. *Everest: The man and the mountain.* Landmark Enterprises, Rancho Cordova, CA. In press.

Snyder, J.P. 1993. *Flattening the earth: Two thousand years of map projections.* Univ. of Chicago Press.

Soler, T. & Hothem, L.D. 1988. Coordinate systems used in geodesy: basic definitions and concepts. *J. Surveying Engineering* 114, No. 2:84–97.

Sprent, A. 1992. VLBI, GPS and Greenhouse. *Aus. Sur.* 37, No. 1:23–32.

Stanbridge, M.J. 1979. Doppler survey—a review of the techniques, systems, equipment, computation and typical results. Conference of Commonwealth Surveyors. Paper B2. Cambridge, UK.

Stansell, T.A. 1983. *The Transit navigation satellite system.* Magnavox, Torrance, CA.

Stead, J. & Holtznagel, S. 1994. AHD heights from GPS using AUSGEOID93. *Aust. Sur.,* March, pp. 21–27.

Stolz, A. et al. 1983a. Australian baselines measured by radio interferometry. *Aust. Sur.* 31, No 8:563–566.

Stolz, A. et al. 1983b. Geodetic surveying with quasar radio interferometry. *Aust. Sur.* 31, No. 5:305–314.

Stolz, A. & Masters, E.G. 1982. Studying the tectonics of Australia by satellite laser ranging. *Aust. Sur.* 31, No. 1:34–44.

Stolz, A. & Masters, E.G. 1983. Satellite laser range measurements of the 3200 km Orroral-Yarragadee baseline. *Aust. Sur.* 31, No. 8:557–562.

Strasser, G. 1974. The toise, the yard and the metre; the struggle for a universal unit of length. 17th Australian Survey Conference, Melbourne.

Thomas, T.L. 1982. The six methods of finding north using a suspended gyroscope. *Sur. Rev.* 26, No. 203:225–235; No. 204:257–272.

Thompson, S.D. *Everyman's guide to satellite navigation.* ARINC Research Corporation.

Treftz, W.H. 1981. An introduction to inertial positioning as applied to control and land surveying. *S. & M.* 41, No. 1:59–67.

Vincenty, T. 1987. Geoid heights for GPS densification. *ACSM Bull.,* Dec., pp. 25–26.

Wells, D.E. et al. 1982. Marine navigation with NAVSTAR/Global positioning system (GPS) today and in the future. *Can. Sur.* 36, No 1:9–26.

Wells, D.E. et al. 1986. *Guide to GPS positioning.* Canadian GPS Associates, Univ. of New Brunswick.

Wells, D. & Kleusberg, A. 1990. GPS: a multipurpose system. *GPS World,* Jan./Feb., pp. 60–63.

Wilson, J.I. & Christie, R.R. 1992. *A new geodetic datum for Great Britain.* The Ordnance Survey Scientific GPS Network SCINET 92. Ordnance Survey.

James (Jim) Smith was educated at Devizes Grammar School and S. W. Essex Technical College (now the University of East London). There he qualified for membership of the Royal Institution of Chartered Surveyors in the Land Survey Division, where he has been active on various committees and the Divisional Council for nearly 30 years. He was closely involved in development of the monthly journal *Land and Minerals Surveying* and its successor *Surveying World*. He at present assists with the contents of *The Civil Engineering Surveyor*.

Internationally he was secretary of Commission 6 (Engineering Surveying) of the FIG (International Federation of Surveyors) for nine years and is now secretary to its ad hoc Commission on the History of Surveying. He has served on British Standards Institution and International Standards Organization working groups on surveying.

He was a principal lecturer in land surveying in the civil engineering department at the University of Portsmouth, Portsmouth, England, for nearly 25 years until taking early retirement in 1990. He is the author of several books and numerous booklets and technical papers, and is former editor of a series of books on Aspects of Modern Surveying. He has recently had published *Capt. R. S. Webb MBE, RFA (1892–1976): from Shropshire to Paarl via Geodesy and Lesotho,* and has *Everest: the Man and the Mountain* awaiting publication.

Index